AIRLINE

DEREGULATION

INTERNATIONAL EXPERIENCES

EDITED BY

KENNETH BUTTON

*Professor of Applied Economics and Transport,
Loughborough University, UK*

NEW YORK UNIVERSITY PRESS
Washington Square, New York

387.71
A2981

First published in the U.S.A. by
NEW YORK UNIVERSITY PRESS
Washington Square
New York, NY 10003

Library of Congress Cataloging-in-Publication-Data
Airline deregulation: international experiences/edited by Kenneth
Button.
    p.   cm.
Includes bibliographical references and index.
ISBN 0-8147-1157-X (cloth):$40.00
    1. Airlines—Deregulation.   2. Aeronautics and state.   I. Button.
Kenneth John.
HE9780.A4 1991   90-49868
387.7′1-dc20   CIP

Manufactured in Great Britain
₸

Contents

v

Contributors

Kenneth Button, Professor of Applied Economics and Transport and Director of the Applied Microeconomics Research Group, Loughborough University, England.

Peter Forsyth, Senior Lecturer in Economics, Australian National University, Australia.

Tae Oum, Professor, Department of Commerce and Business Administration, University of British Columbia, Canada.

Donald Pickrell, Economist, US Department of Transportation (Systems Center), USA.

William Stanbury, UPS Foundation Professor of Regulation and Competition Policy, University of British Columbia, Canada.

Dennis Swann, Professor of Economics, Loughborough University, England.

Michael Tretheway, Associate Professor, Department of Commerce and Business Administration, University of British Columbia, Canada.

List of Figures

List of Tables

CHAPTER 1

Introduction

Kenneth Button

The past two decades have seen remarkable changes in the way that economic regulation has been viewed, for example, see Swann (1988) and the papers in Button and Swann (1989a). Countries have varied over time in the extent and the ways in which they have intervened in economic markets, particularly in terms of price and entry controls, but the tradition of regulation is well established. What has happened in recent years has been a liberalization of attitude and something of a withdrawal of the state from this interventionist role. In North America this has been viewed as a period of 'deregulation' although the UK terminology of 'regulatory reform' is in many ways more appropriate. While many economic regulations have been removed in some cases there have been tightenings of controls of quality factors related, for instance, to safety. Equally, in countries such as the UK and France where there has been privatisation of formerly state-owned industries, economic controls have actually been reinforced to replace the direct control of ownership formerly exercised over these large-scale undertakings. There has also been, in some instances, a tightening of broader industrial policy and labour protection laws as individual sectors have been liberalized – the new mergers policy of the European Communities (EC) from 1989 is an example. The gradual 'greening' of attitudes towards industry has also brought forth new regulations governing the ways in which we treat the natural environment.

While these changes extend across many spheres of economic activity they have been particularly pronounced in the context of transport. The long-standing tradition in transport had been one of market intervention by government to regulate entry and/or price with the intent, on the one hand, of protecting consumers, third parties and those working in the industry together with, on the other hand, the achievement of social objectives such as service to remote communities and the integration of spatially disparate markets. Indeed, in many countries, and especially on continental Europe, transport has traditionally been seen as an input into a wider socio-political-economic process embracing regional, social and industrial policies. As such it has been manipulated to achieve aims which transcend issues of simple efficiency in transport supply.

With one or two notable exceptions (e.g. the deregulation of the UK trucking industry in 1968 and partial deregulation of the railways in 1962), the controls

1

which had been built up from the nineteenth century and developed and rationalized in the inter-war period continued to dominate into the 1970s. The 1930s had been a particularly active period for regulators (Button & Gillingwater, 1986) and much of the change which has occurred in recent years effectively rolled back the legislation of that time. A considerable part of this earlier legislation can be explained in terms of society trying to come to terms with new modes of transport (especially the automobile and aeroplane) which not only offered new technical challenges but also began to have a very considerable effect on the way people lived. In some ways, therefore, one could view the legislation as a form of social engineering. Equally, however, transport has associated with it a wide range of externalities – it is dangerous, pollutive and affects a wide range of other economic activities ranging from residential land use to international trade patterns. Issues of overall social efficiency were, therefore, also at the forefront of the debates of the time.

From the mid-1970s economic liberalization has spread through transport markets with supplying industries being freed from price and entry constraints and privatisation taking effect in many sectors. The reasons behind the change vary from country to country and sector to sector. The USA is often seen as the initiator of most of the reforms with a combination of demonstration effects and bandwagon effects, together with some clear spillovers into international markets, causing other countries to imitate. Of course one could debate the claims of the US to be at the forefront of change in transport policy – the UK in particular may feel it has some precedence – but it was certainly the experiences of the US which have received the greatest attention.

During a period from 1976 to the early 1980s the Airline Deregulation Act (1978), the Staggers Rail Act (1980), the Motor Carrier Act (1980), and the Bus Regulatory Reform Act (1982) essentially liberalized inter-state transport in the USA with knock-on effects rippling through to intra-state regulation. Following lags, similar pictures emerge for most other industrialized countries although the timing and intensity of change has varied enormously. There are many reasons for this diversity of pace. In part it is explained by the different starting points from which liberalization began, but equally one must consider the nature of the political systems involved – some simply do not have the mechanisms for rapid change, the political philosophy of the different countries, the direct links with other, liberalized transport networks, and the degree to which regulations were functioning effectively.

The aim here is not to delve in great detail into the causes of change in various countries or why the outcomes in each are unique. These are issues which are brought out in the contributions to the book. Indeed, one of the objectives of this collection is to consider whether our experiences of market liberalization reveal any common threads. In particular, whether they reveal any broad, universal indications of how underlying transport markets function; how management responds to new stimuli; the degree to which transport users and third parties really need protecting from the antics of supplying industries; and the nature of transition processes from regulation to liberalization.

These are not just issues of academic interest. As we have said, liberalization has been pursued at different rates in different countries and for different modes and there has clearly been something of a bandwagon effect in operation. Separation of the general effects from the contextual, therefore, seems important for on-going policy development. There are, in particular, possible lessons, for those still framing liberalizing codes, to be learned from countries which have set the pace in reform. Ideally, in this context, one would seek to cover all modes in such an analysis, if for no other reason than that modal substitution can influence the behaviour of a particular sector, but pragmatism leads us to focus primarily on aviation.

The choice is not a random one. First, aviation is an international industry and therefore changes are of international importance (Kasper, 1988; Doganis, 1989). Second, from an analytical perspective, we now have considerable experience of the effects of liberalization from the happenings in the USA since the passing of the Airline Deregulation Act (Morrison, 1989; Kahn, 1988). This, in particular, helps our understanding of the dynamics of the post-liberalization phase and the reactions of supplying airlines to it. From a more topical perspective we have the important debates and developments which are going on within the EC as a common Community approach to aviation evolves (Button & Swann, 1989b; Argyris, 1989; McGowen & Seabright, 1989). There are also on-going developments in other countries. Australia, for instance, is radically modifying its domestic aviation strategy as it moves away from the long-established two-airline policy (Kirby, 1981). Conversely, Canada is struggling with what has essentially become a two-airline domestic industry since deregulation (Gillen *et al.*, 1988; Barone *et al.*, 1986).

While the contributions to this volume are not designed to be overly theoretical, they do address some interesting theoretical, as well as policy-orientated, issues. They, for example, shed some light on the on-going debate about the underlying nature of the aviation market. The up-rising of 'contestability theory' in the late 1970s led many to believe that ultra-free market entry and exit would ensure efficiency in the sector with maximum benefits being conferred on users. The empirical evidence has not always supported this rather idealized view of the outcome and now we find those who feel the market could more usefully be described as: 'imperfectly contestable' (Morrison & Winston, 1987; Levine, 1987), 'workably competitive' (Keeler, 1990), broadly oligopolistic (Kahn, 1988), etc. This in turn leads onto questions of optimal regulatory regimes and the difficult trade-offs which must be made between the implications of markets failures and those which often accompany government intervention and regulation.

The chapters in this volume cover these main themes in the context of four substantive case studies – of the USA, Australia, the UK and Canada. Each of these sets out the development of aviation policy in the subject country, the specific reasons behind subsequent moves towards liberalization and, where appropriate, comments on the impacts of these changes. There is also some forward thinking as to where the changes are likely to lead in the longer term.

References

Argyris, N. (1989) 'The EEC rules of competition and the air transport sector', *Common Market Law Review*, **26**, 5–32.

Barone, S. S., Javidan, M., Reschenthaler, G. B. & Kraft, D. J. H. (1986) 'Deregulation in the Canadian airline industry: is there room for a large regional carrier?', *Logistics and Transportation Review*, **22**, 421–48.

Button, K. J. & Gillingwater, D. (1986) *Future transport policy*, Routledge: London.

Button, K. J. & Swann, D. (eds) (1989a) *The age of regulatory reform*, Oxford: Clarendon.

Button, K. J. & Swann, D. (1989b) 'European Community airlines – deregulation and its problems' *Journal of Common Market Studies*, **27**, 259–82.

Doganis, R. (1989) 'Regulatory changes in international air transport'. In Button, K. J. & Swann, D. (eds), *The age of regulatory reform*, Oxford: Clarendon.

European Communities Commission (1985) *Completing the Common Market*, COM (85), 310 Final, Brussels, European Commission.

Gillen, D. W., Stanbury, W. T. & Tretheway, M. W. (1988) 'Duopoly in Canada's airline industry: consequences and policy issues', *Canadian Public Policy*, **14(1)**, 15–31.

Kahn, A. E. (1988) 'Surprises of airline deregulation' *American Economic Review, Papers and Proceedings*, **78**, 316–22.

Kasper, D. M. (1988) *Deregulation and globalization: liberalizing international trade in air services*, Cambridge, Mass: American Enterprise Institute/Ballinger.

Keeler, T. E. (1990) 'Airline deregulation and market performance: the economic basis for regulatory reform and lessons from the US experience. In Banister, D. & Button, K. J. (eds), *Transport in a free market economy*, London: Macmillan.

Kirby, M. G. (1981) *Domestic airline regulation: the Australian debate*, Sydney: Centre for Independent Studies.

Levine, M. (1987) 'Airline competition in deregulated markets: theory, firm strategy, and public policy' *Yale Journal on Regulation* **29**, 393–494.

McGowan, F. & Seabright, P. (1989) 'Deregulating European airlines', *Economic Policy*, **9**, 283–344.

Morrison, S. (1989) 'US domestic aviation'. In Button, K. J. & Swann, D. (eds), *The age of regulatory reform*, Oxford: Clarendon.

Morrison, S. A. & Winston, C., (1987) 'Empirical implications and tests of the contestability hypothesis', *Journal of Law and Economics*, **30**, 53–66.

Swann, D. (1988) *The retreat of the state – deregulation and privatization in the UK and the USA*, Hemel Hempstead: Harvester-Wheatsheaf.

CHAPTER 2

The regulation and deregulation of US airlines

Donald Pickrell

2.1 Introduction

Throughout its history, regulation of airline service in the United States sought to suppress the threat of competition. Although regulators' motives for doing so varied over their fifty-year reign, from promoting nationwide air service to stabilizing the fledgling industry and ultimately to protecting the financial interests of individual carriers, their actions systematically foreclosed entry into the industry while progressively narrowing the opportunities for competition within the industry. In an abrupt reversal of its own historical policies, the industry's regulatory body took the first steps to relax its tight controls over fares and service during the mid-1970s.

Emboldened by the results of this experiment with 'deregulation' (as it came to be called), the Congress codified these and even more sweeping reforms – including eventual elimination of the regulatory agency itself – in the Airline Deregulation Act of 1978. The industry has become dramatically more competitive in the slightly more than a decade since passage of the Act, and this intensified competition has produced new travel opportunities, improved service, and lower fares for the vast majority of the nation's airline passengers. Nevertheless, certain developments in the deregulated industry have surprised even its most insightful observers, and some potential threats to the continued vigour of competition remain to be resolved.

The origins of US airline regulation

While formal regulation of passenger service dates from passage of the Civil Aeronautics Act in 1938, both the structure of the industry and the pattern of regulation were established by federal regulation of airmail service beginning twenty years earlier. Commercial air transportation began with the award of airmail contracts to private air carriers in 1925 by the US Post Office, which had operated airmail service using pilots and aircraft supplied by the Army since 1918. During the following decade, the Post Office used these awards to expand the nation's route system, while consciously avoiding competition on

5

established routes and promoting the development of a few large carriers. It also used the structure of subsidy payments to encourage carriers to jointly provide passenger service, as well as to acquire aircraft specifically suited to this new purpose (Meyer & Oster, 1981).

After irregularities in subsidy awards led to an abortive attempt to return to US Army operation of the airmail service, new route awards were made and authority to set rates was transferred to a long-established regulatory agency, the Interstate Commerce Commission (ICC). Financial difficulties experienced by many airlines as a result of the ICC's policies (which encouraged carriers to accept mail at rates below cost by awarding new routes to those who did so) threatened the industry's stability, and in combination with the growing importance of passenger services led to passage of the 1938 Civil Aeronautics Act (Kaplan, 1986). This legislation established the Civil Aeronautics Agency (which was reorganized in 1940 to become the Civic Aeronautics Board, or CAB), which it provided with a mandate order to do so, the Act authorized the agency to award individual routes to carriers, specify the fares they were allowed to charge, and ensure safe airline operating practices.[1]

The newly-created regulatory agency was chartered to allow competition only to the extent necessary to do so, while explicitly avoiding 'unfair or destructive' competitive practices. As with much other economic regulation adopted during the period, this mandate reflected the widespread public scepticism of the consequences of unchecked competition – particularly the threat of potentially 'excessive' competition – that had arisen in the catastrophic economic conditions of the Great Depression. The CAA's first important action, granting permanent authority for existing airlines to serve routes over which they operated at the time of the Act's passage, was not only important of itself but presaged its cautious regard for the spectre of competition over most of the subsequent four decades. Thus despite a remarkable succession of innovations in aircraft technology, and the phenomenal growth and development of the industry they fostered, the Board consistently resisted new carriers' efforts to enter the industry, while progressively broadening its controls over the activities of existing airlines.

The industry's development under regulation

After 'grandfathering' the route authority of the 16 airlines holding it in 1938, the CAB precluded entry into service by new carriers, refusing nearly one hundred petitions for new service between its establishment in 1938 and the onset of deregulation in the mid-1970s (Kaplan, 1986). It also carefully restricted awards to established carriers of authority to serve new routes, denying most requests and granting others only after protracted hearings in which petitioners were required to demonstrate how the proposed service would further the public interest without causing financial harm to any incumbent carrier. While most of its grants of new route authority were apparently intended to strengthen financially weak carriers by authorizing them to carry profitable traffic, even this 'levelling' policy was insufficient to prevent consolidation of the industry from its original 16 members to the 11 that existed when the Board began to deregulate the industry.[2]

The CAB's major service initiative was to permit a group of 'local service carriers' to begin subsidized feeder service connecting small communities with the cities served by the original 'trunkline' carriers during the early 1950s, although the Board prohibited them from competing with established airlines (and even unsuccessfully sought to eliminate them after deeming its 'experiment' a failure). The structure of subsidy payments and the availability of guaranteed loans subsequently encouraged these airlines to replace their original fleets of surplus military aircraft with new and more costly models, however, while political pressures to extend the subsidized service they provided were difficult for the Board to resist (Eads, 1972). Subsidy payments to the local service airlines grew rapidly as a result, forcing the Board reluctantly to begin awarding them more profitable, longer-haul routes (in the hope of offsetting their losses on feeder routes) on which they often competed with the trunk carriers. By subsequently encouraging local service carriers to acquire jet aircraft to serve these new routes, the Board promoted their evolution into major jet airlines serving regional route networks, many of which became important competitors in the environment later fostered by deregulation.

Despite the CAB's efforts to restrain competition, the industry grew extremely rapidly in the postwar era, carrying well over ten times as many passengers during 1970 as it had 20 years earlier. This growth was fuelled by the sustained economic growth the nation experienced during the postwar period, coupled with the rapid pace of innovation in aircraft design, which reduced carriers' costs sharply and thus allowed the Board to maintain stable fares for almost two decades.[3] The new aircraft technologies also reduced airlines' costs for providing long-haul service by more than they reduced those for short flights, so that by failing to adjust fares to reflect this changing cost structure, the CAB was able to cross-subsidize the development of short-haul service with profits earned on longer routes. The Board also took advantage of the rapid decline in carriers' costs fostered by the introduction of jet aircraft during the 1960s to permit airlines to offer discount fares, which contributed to the rapid growth in air travel.

Stresses on the regulatory regime

By 1970, however, the industry's conversion to jet aircraft was virtually complete, and carriers' costs began for the first time to reflect rising prices for labour, fuel, and other operating inputs. The ensuing application for fare increases by airlines facing severe pressures on their profits ultimately forced the CAB to re-examine its fare-setting policies as part of its Domestic Passenger Fare Investigation (DPFI). Although the DPFI led the Board to correct certain distortions arising from its fare policies, it adopted a rigid formula that set fares strictly as a function of distance travelled (which maintained fares above costs on long routes and below costs on shorter flights), and rescinded all discount fares as 'discriminatory'. During the course of the DPFI, moreover, the Board adopted two particularly controversial measures intended to protect the industry's profitability, imposing a moratorium on its award of new route authority, and sanctioning negotiations among the largest carriers to mutually limit flights in their busiest markets.

In the midst of the furore created by the capacity limitation agreements arrived at by carriers acting under its authority, the CAB also moved to adopt minimum fares for charter airlines, which had begun to make competitive inroads into the regulated carriers' traffic on transcontinental routes. Shortly after that it approved a general fare increase which, while relatively small, combined with two other recent increases and the elimination of discount fares recommended as part of the DPFI, to raise average fares nearly 20% within one year (Meyer & Oster, 1981). These actions spurred a powerful Congressional committee to conduct well-publicized hearings on the effects of its recently adopted policies to limit fare and service competition, during which several witnesses testified that CAB regulatory policies resulted in excessive fares, as well as carriers' acquisition and operation of unnecessary aircraft and flights. Even the CAB itself – within which many of these measures had been controversial – convened a special task force to study potential revisions of its regulatory procedures, which recommended that the Board rescind the capacity limitation agreements and relax its rigorous control over fares and airlines' entry into new markets.

The CAB's initial steps toward deregulation

In the Spring of 1975, newly-appointed CAB chairman John Robson proposed an experiment that would permit new and existing airlines to begin or end service on selected routes without prior CAB approval, a dramatic departure from the progressively tighter regulatory policies advocated by his immediate predecessors. Although this predictably controversial proposal was never adopted, later that year the Board authorized new competition on a number of routes for the first time since 1970, including some where existing service was satisfactory. More important, during 1975 and 1976 the CAB relaxed a number of restrictions it had previously imposed on charter carriers, thereby allowing them to operate frequent low-fare service in direct competition with regulated carriers. It also allowed airlines to reinstate limited discount fares, and awarded new route authority to various applicants who proposed to operate trans-continental service at considerably lower fares than were then in effect.

In response to the threat of increased competition from charter carriers and low-fare scheduled airlines, American Airlines applied for and was granted authority to offer discounts of up to 45% from coach fares (subject to certain restrictions on their use) for travel in the transcontinental markets it served. When they were first offered in the spring of 1977, American's 'Super Saver' fares resulted in traffic increases exceeding 60% on some routes, leading the Board – under the leadership of newly-appointed chairman Alfred Kahn – to enthusiastically endorse other carriers' subsequent requests to offer similar discounts. By the following spring, the regulated airlines were offering discount fares (still subject to advance-purchase and minimum stay restrictions) on virtually all their routes, and the CAB proposed that carriers be allowed to reduce fares as much as 70% below the level called for by the fare-setting formula adopted during the DPFI without prior approval. Further, after experimenting throughout the previous year with a policy that permitted all applicants seeking a new route to simultaneously begin serving it, the Board proposed to allow carriers to begin serving new routes (and to withdraw from others) without conducting its usual case-by-case review of their applications (Kaplan, 1986).

The enactment of formal deregulation

Amid prosperity in the airline industry unseen during the previous 10 years, the US Congress enacted the Airline Deregulation Act in October of 1978. While the Act largely codified the fare flexibility and more liberal route award policies that had been developed by the CAB over the previous year, it went beyond them by terminating the Board's jurisdiction over carriers' route networks in three years, and phasing out its authority to set fares over a five-year period. For the first time, it made 'commuter' airlines – which operated small propeller-driven aircraft and had been exempted from CAB regulation of their fares and route authority – eligible to receive subsidies for serving small communities. The Act also extended subsidies to underwrite continued service to communities where regulated carriers used their newly-granted freedom to abandon service.

Sounding a note of finality, the Deregulation Act also required Congress to decide whether to permanently eliminate the CAB after the demise of its regulatory authority; in 1984 it did so, transferring the Board's remaining jurisdiction over airline mergers, consumer complaints, and other matters to the Department of Transportation.[4] Thus the US airline industry was rapidly moved from an environment of progressively tighter controls over a growing number of its activities to one of virtually unfettered ability for its members to select the markets they served, adjust the frequency of service they offered, and set the fares they charged. Perhaps the most remarkable aspect of this unprecedented reversal in government regulatory policy was that *both* the final tightening of regulatory controls and their subsequent elimination were presided over by the same agency, acting almost completely on its own initiative, within the span of a few years. The weight of government restrictions on virtually every conceivable tactic airline managements might have used to wage competition, which had accumulated over nearly 50 years, had thus been lightened with a suddenness that – at least from the perspective of the slightly more than a decade that has elapsed since – certainly seems dramatic.

2.2 Factors influencing the response to deregulation

The US airline industry's rapid transformation during its little more than 10 years of experience with deregulation has been the product of a number of important forces. These include the apparently substantial operational and marketing advantages associated with large hub-and-spoke route systems, continuing growth in the demand for intercity travel, and sharp fluctuations in the costs of important inputs used by air carriers. More recently, difficulties in co-ordinating pricing and investment policies governing the provision of public infrastructure inputs used to produce air transportation service – primarily airport facilities and airspace capacity – appear to have exerted an increasingly important effect on airline service patterns and fare levels.

Although some of these forces were at work well before the movement toward deregulation began, the industry's evolution in response to them was slowed by the CAB's exercise of its fare-setting and route award authority. Perhaps as a result, these forces appeared to receive relatively little attention during the pre-deregulation debate, thus causing the pace and extent of certain developments they have spawned to surprise even many experienced airline industry observers. Whatever the exact source of inattention to these forces, it is now apparent that they represent the dominant influence on the industry's continued evolution. Further, insofar as they are present in other situations where deregulation is contemplated – and virtually all appear to be global rather than local phenomena – they can be expected to shape the response of those industries as historical controls over firms' pricing and service prerogatives are relaxed or eliminated.

Advantages of large hub-and-spoke route networks

Larger airline networks organized around major route hubs appear to provide important advantages to their operators in both producing and marketing air travel services. On the production side, these networks facilitate the use of larger aircraft, which offer lower unit (i.e. per seat or seat-mile) operating costs on most routes. Hub-oriented networks also allow airport facilities to be more intensively utilized, thus reducing airlines' average costs for leasing and operating them. On the marketing side, routing flights through a hub facilitates more frequent departures to a large number of cities, thereby making service more attractive to travellers destined for each of them. In addition, large hubs appear to allow carriers who operate them to charge higher fares than competition would otherwise permit, and also seem to enhance the effectiveness of recent marketing innovations (such as frequent-flyer bonuses) in raising load factors on their operators' flights.

Economies from operating larger aircraft

Both engineering-economic and econometric studies provide evidence that above some minimum route length, aircraft operating costs per seat-mile decline as larger aircraft are used (Viton, 1986). Thus as the number of passengers travelling on a route connecting two cities increases, airlines can reduce their per-passenger aircraft operating expenses on the route by substituting progressively larger aircraft without experiencing a decline in load factors. Because rising demand also allows the frequency of scheduled departures to be maintained, the resulting service will remain equally desirable from passengers' viewpoint. Recent studies (Caves *et al.*, 1983; Gillen *et al.*, 1985) suggest that many airline routes in North America may be subject to such average cost reductions, or 'economies of density', resulting from increasing passenger volumes.

One way carriers can increase the number of passengers travelling on flights between two cities without undertaking measures (such as lower fares) to increase the volume of travel is by routing flights between the two cities through a third point (rather than non-stop). Traffic on each resulting flight will represent the sum of passengers travelling between each of the two city-pairs that has the hub city as one of its end points. For example, suppose an airline provides service in each of three city-pair markets, A to B, A to C, and B to C (ignoring reverse flows and service for simplicity). The carrier can combine the travel volume between A and C with that between A and B by routing its flights between cities A and C via B, which functions as the network's hub. By doing so, passenger volumes on the A–B and B–C routes will increase sufficiently to support operation of larger aircraft on both.[5]

Of course, carriers will incur some additional costs by routing flights between any two 'spoke' cities through a hub, including higher aircraft operating expenses for the extra takeoff and landing and any additional distance travelled, as well as some additional costs of handling the increased flow of passengers through the network's hub city. As long as these additional costs do not offset the savings realized by operating larger aircraft, however, an airline's total costs of operating the simple service network described in this example will decline when it is configured around a central hub, as will the resulting average per-passenger costs of serving each pair of cities comprising the network.[6]

Implications for network-wide economies

Over a network entailing service in a large number of city-pair markets, the economies of density made possible by aggregating passenger flows on flights to and from a hub can be quite large. This occurs because the number of city-pair passenger flows that can be aggregated on any spoke of such a system increased directly with the number of cities comprising the network. For example, traffic on any one route in a network consisting of 10 spoke cities and one hub will represent the sum of travel between a spoke city and each of 10 other cities, or 20 city-pairs. If the number of spoke cities served as part of the network is doubled to 20, traffic on any route will represent the sum of travel in 20 separate city-pair markets.

Channelling traffic in numerous city-pair markets through a single hub in this manner not only allows a carrier to realize the substantial economies of operating larger aircraft, but also allows it to schedule more frequent departures on each route. More frequent services can thus be offered by the carrier to offset the disadvantages to passengers of the additional elapsed time and inconvenience entailed in travelling through the hub. In a large network, these disadvantages can be further reduced by scheduling flights from each of the spoke cities to converge on the hub at approximately the same time, and to depart for their destinations a short time later, thereby minimizing the delays experienced by passengers in making transfers.[7]

In combination with factors such as shorter walking distances between passenger gates and a reduced probability of missed flight connections or lost baggage, the resulting convenience of such 'online' connections – those between flights operated by the same carrier – also causes passengers to strongly prefer them (Carlton, *et al.*, 1980). This preference for online connections means that a large hub-and-spoke network operated by a single carrier will enjoy a competitive advantage over an equally large network of service relying on connections between flights operated by two or more carriers, unless those carriers are able to co-ordinate their activities so as to duplicate the conveniences of online connections.[8]

The economies offered by integration of a large-scale route network through a hub represent an example of economies of scope, since by serving a large number of city-pair markets through a hub, a carrier in effect provides many different 'products'. Similarly, airlines realize operating economies from serving classes of travel demand that have different characteristics – principally business and vacation travel – using a single network of flights, another example of the economies of scope attainable from operating a large hub-and-spoke route network. Airlines may also be able to realize further economies in serving a complex network of city-pair travel demands by operating multiple hubs at different cities, as most large US carriers now do. Further economies can arise in a multi-hub network for two reasons. First, by serving many of the same cities from both hubs, a carrier can lower the average per-passenger costs it incurs in providing and operating facilities at each spoke city, since the increase in those costs will generally not match the resulting increase in passenger volume. Second, establishing an additional hub offers important operational advantages, such as the potential to achieve more efficient scheduling and thus fuller utilization of aircraft and flight crews (US Congressional Budget Office, 1988).

Marketing advantages stemming from network size
For a variety of reasons, an airline that offers the most service inbetween two cities tends to attract a disproportionate share of passengers travelling between them (Fruhan, 1971; Bailey, *et al.*, 1985; Levine, 1987). Anticipating that the airline operating the largest number of flights is most likely to offer one near

their respective desired departure and return times, most travellers are likely to contact that carrier first (or to ask a travel agent to examine its schedule first). Further, flights operated by the airline offering the most frequent departures represent better substitutes for one another, in the event that a traveller is unable to obtain a seat at the most desirable departure time, or is forced to reschedule in response to a missed flight or a change in travel plans. Thus a carrier able to offer frequent flights in a large number of markets is likely to enjoy an advantage over smaller competitors in marketing its service, and as discussed previously, the most efficient way for a carrier to do so is by operating a large hub-and-spoke route network.

In addition, operating a large hub appears to allow airlines to charge higher fares than would otherwise be possible on routes emanating from the hub, while minimizing the usual increased risk of attracting entry by potential competitors (Borenstein, 1988). This appears to occur partly because the existence of a large hub operation raises the minimum scope and frequency of services with which another airline would have enter the market in order to represent a viable competitive threat. Both the total costs to potential entrants of providing this minimum level of service and the fraction of those costs that are 'sunk', or non-recoverable in the event an entrant subsequently decides to withdraw, are likely to increase with the size of the incumbent carrier's existing network. In addition, the incumbent airline may exercise sufficient control over passenger and baggage handling facilities at its major hub to directly restrict entry by potential competitors, simply by refusing or blocking access to these facilities by potential entrants.

A large hub also seems to increase the effectiveness of certain marketing innovations developed during the decade of deregulated competition in raising load factors on an airline's flights. Since airlines' costs per passenger decline rapidly as the fraction of seats on its flights that are filled increases, operating a large hub-and-spoke route system can thus improve a carrier's profitability. For example, travellers can most rapidly accumulate frequent-flyer bonuses (which allow free travel after a passenger purchases some minimum amount of service) on the airline operating the largest number of flights to and from the cities where they live. Because such bonuses are also most valuable when they can be redeemed for travel to a wide range of destinations, the opportunity to accumulate them is most attractive to travellers residing in a city when it is offered by a carrier operating a large hub there.

Another recent marketing innovation is the payment by airlines of bonus or incentive commissions (often called 'commission overrides') to travel agents who book more than some specified number of passengers for travel on their flights. The number of bookings above which an airline pays such a bonus can simply be some estimate of the number the airline would normally expect to receive from that agent, or it can be based on a marketing target the airline sets for a new service it has introduced. These bonuses usually take the form of an increase in the commission rate paid by an airline on the dollar value of some fraction of the tickets for travel on its flights that is sold by an agency. Although overrides often apply to all sales by a travel agency with which an airline establishes an incentive agreement, they can also apply only to bookings above the expected number, or to tickets sold for travel on specific routes or even individual flights.

Although commission overrides are not necessarily a more effective marketing strategy for large airlines than for smaller ones, paying overrides may actually be less costly on a per-passenger basis for large airlines than for their smaller competitors. This is because larger carriers with established reputations and more frequent flights may only find it necessary to pay a higher commission rate on a small fraction of their passenger bookings – such as those above each agency's expected level – in order to influence travel agents to recommend more strongly their flights to customers. In contrast, smaller carriers offering less frequent service or having newly entered a market may find it necessary to pay larger bonuses, or to make a larger fraction of an agency's bookings eligible for those bonuses, in order to persuade the agency to promote their services more vigorously.

Finally, large airlines have been those most easily able to undertake the substantial investments necessary to develop and market computer reservation systems (CRSs) to travel agents. Although their ownership is now more diversified, each of the five airline-owned CRSs now in use by travel agents was initially developed by one of the five largest pre-deregulation airlines. Each system initially displayed flights operated by its airline owner more prominently than those of its competitors, and because travel agents tend to suggest flights to customers in the order that they are displayed, this practice increased the likelihood that customers would use flights operated by each system's airline owner rather than those of its competitors. While using carriers' identities to rank flights for display in CRSs was subsequently prohibited by the CAB, travel agents' use of these systems apparently continues to produce a 'halo' effect, whereby agents tend to book a larger share of passengers on flights operated by the airline owners of the systems they employ than would be expected on the basis of those airline's competitive presence in the local market (US Department of Transportation, 1988).

As a result, the largest airlines have been able to use these systems to enhance their competitive positions at the expense of smaller carriers, by inducing agents who use their CRSs to divert passengers who would otherwise have travelled on competing airlines to their own flights. At the same time, because travel agents equipped with CRSs have become a vital marketing outlet for virtually all airlines, it appears that CRS-owning airlines have been able to charge their competitors fees for the use of these systems that substantially exceed the costs they incur in processing and storing passenger reservations. This ability to charge booking fees that exceed the costs of providing reservation services has raised the costs to smaller airlines that do not own CRSs, of marketing and distributing their services, thus accentuating the competitive advantage enjoyed by larger airlines over their smaller rivals.

Increasing demand for airline travel

Since passage of the Airline Deregulation Act in October of 1978, the number of domestic passengers carried by the nation's airlines has grown by 62% while the number of passenger-miles travelled aboard their domestic flights has risen by 79%. Although some of this dramatic increase in air travel is attributable to reductions in the overall level and changes in the structure of airfares fostered by deregulation, much of it undoubtedly is the result of continuing long-term growth in nationwide demand for air travel. This growth has in turn been the product of changes in business activity and household circumstances that have contributed to growing demand for intercity travel in general, and for the particular characteristics – speed, comfort, and safety – featured by airline service. Despite the interruption of these forces by the prolonged economic recession of the early 1980s, the demand for air travel appears to have expanded robustly since deregulation of the industry began.

The demand for business-related airline travel has been spurred by both continuing growth in the level of business activity (although this was temporarily interrupted by the recession of the early 1980s), and an increase in the fraction of business activity that is responsible for the largest share of business-related air travel. During the slightly more than a decade since the US domestic airline industry was deregulated, total non-agricultural employment – which provides a broad index of the level of business activity nationwide – has risen more than 25%. More important, employment in those industries that rely most heavily on face-to-face personal contacts to transact business, and thus generate a disproportionate share of business demand for airline travel, increased nearly 60% during the same period.[9]

Most determinants of the demand for vacation and personal business travel by individuals and households also continued to change over the last decade in ways that contributed to growing use of airline service.[10] Most important, real (or inflation-adjusted) per capita personal income of the US population rose by slightly more than 25% between 1978 and 1988, thus increasing individuals' willingness to pay for the superior speed and comfort afforded by airline travel. At the same time, the average number of hours per week that Americans spent at their jobs continued its historical decline, thereby making more time available for conducting personal business, pursuing recreational opportunities, and other non-work activities that often entail intercity travel. Finally, the average size of US households continued to decline slowly during the 1980s, while the fraction of those households that consist of a single individual or of unrelated individuals continued to grow. Because intercity trips made by smaller travelling parties are more likely to be made by air (Meyer & Oster, 1987), these continuing changes in the composition of US families may also have contributed to growth demand for airline travel during the period since deregulation began.

Supply and access to airport and airway facilities

Airlines employ two important governmentally-supplied inputs in producing air travel services: airport capacity, and the navigation and aircraft control services provided by the national airspace system. These two inputs are provided by two co-ordinated but separate systems of nationwide aviation infrastructure, the system of commercial airports and the airspace system that control aircraft flying from one to another. In the US, the federal government owns and operates the airspace system, which consists of a large number of facilities and personnel located at airports and in a variety of other places, while commercial airports – each of which includes runways, taxiways, boarding gates, and passenger terminal buildings, are individually owned and operated by local (or in a few cases, state) government agencies.

During the period of deregulated airline competition, demand by airlines for both airport capacity and air traffic control services has increased dramatically. Between 1978 and 1988, the US airline industry's fleet of jet aircraft grew by nearly 70%; during that same period, the number of takeoffs and landings at the nation's airports – and thus the number of flights handled by the nation's airways system – rose by 22%.[11] During this period, the average length of those flights also grew by nearly one-third, with the typical flight thus requiring considerably more en-route navigation and traffic control services. Finally, the number of passengers boarding flights at the nation's commercial airports, another indicator of the demand for airport services, increased by 63% over the decade.

During this period of rapidly growing demand for airport capacity, local government agencies responsible for commercial airports have experienced considerable difficulty in making investments or operational improvements that would expand the capacity of the airport and airway systems. Thus the runway and terminal capacity of many US airports has not increased sufficiently to accommodate growing demand by airlines for passenger-handling facilities, boarding gates, or departures and arrivals by their aircraft. Although airports charge fees to airlines for aircraft landings, these are usually insufficient to prevent recurring congestion of their taxiways and surrounding airspace by aircraft seeking to depart from or land at large airports. In addition, the inability of many airports to invest in expanded passenger terminal areas and additional boarding gates has often made it more difficult for airlines seeking to provide new or expanded service to obtain access to facilities that are necessary for them to operate at a scale sufficient to become viable competitors.

At the same time, the federal government has been unable to expand substantially the capacity of the airspace system to accommodate airlines' growing demand for the navigation and air traffic control services it provides. A strike by air traffic controllers during 1981, which led to the dismissal of nearly three-quarters of the controller work force, resulted in severe limits on air traffic at the nation's largest airports that remained in place until 1983, and the controller workforce has only recently regained its original size. In addition, much of the equipment comprising the airspace system is aged and increasingly difficult to maintain, while the federal government's programme of investments in modernizing and replacing the system's infrastructure has been extensively delayed by a combination of technological and political problems. Thus the airspace system seems likely to continue to be hampered by limited capacity and unreliability for the immediate future. Since access to services provided by the system is provided on a first-come, first-served basis, the result is likely to be occasional severe congestion of airspace facilities during periods of peak demand or adverse weather, which will continue to require temporary restrictions on airlines' operations.

2.3 Major developments in the US airline industry

In response to the variety of factors influencing its development, the US domestic airline industry has been radically transformed since the passage of the Airline Deregulation Act. Although of most of the major post-deregulation developments began to affect the industry well before even the CAB began to loosen its controls over airline service and fares, their pace has accelerated dramatically in the more intensely competitive environment fostered by deregulation. The most obvious and far-reaching of these changes in the industry's structure include the rapid expansion and subsequent (equally rapid) consolidation of the number of airlines comprising the industry, these carriers' accelerated development of large hub-and-spoke route networks, and their growing reliance on independent travel agents equipped with airline-owned computer reservation systems for marketing and distributing air travel services.

The industry's expansion and subsequent consolidation

In the years immediately following passage of the 1978 Deregulation Act, many of the former local service airlines grew rapidly and began to provide serious competition, especially for the smaller trunk airlines. As Table 2.1 indicates, they were joined in this role during the early 1980s by a host of new competitors, including former intra-state carriers expanding into inter-state service, new entrants into the industry, and growing commuter airlines that began operating jet aircraft on longer routes. Thus by 1984, the number of jet-equipped airlines providing sufficiently frequent service to compete viably in national or important regional markets had nearly doubled from the 19 in operation immediately prior to deregulation.

Between 1984 and 1986, however, a combination of financial failures, mergers – some motivated by financial distress – and acquisitions rapidly reduced the number of viable competitors in the industry. Thus by 1988, only approximately 10 carriers of sufficient size to represent important regional or national competitors remained in operation. Although this consolidation

Table 2.1   Airlines providing inter-state jet service during the deregulation era.

Origin and name	Began service *	Date	Status
Trunk carriers (11)			
American	pre-1978	1989	1st ranking carrier**
Braniff	pre-1978	1982	Ceased operation due to bankruptcy
		1984	Resumed limited service
		1989	Ceased operation due to bankruptcy
Continental	pre-1978	1989	6th ranking carrier (under Texas Air Corp)
Delta	pre-1978	1989	2nd ranking carrier
Eastern	pre-1978	1989	Declared bankruptcy; conducting limited operations (under Texas Air Corp)
National	pre-1978	1980	Acquired by Pan Am
Northwest	pre-1978	1989	5th ranking carrier
Pan Am	pre-1978	1989	12th ranking carrier
TWA	pre-1978	1989	8th ranking carrier
United	pre-1978	1989	3rd ranking carrier
Western	pre-1978	1986	Acquired by Delta
Local service carriers (8)			
Frontier	pre-1978	1985	Acquired by People Express
Hughes Airwest	pre-1978	1980	Acquired by Republic
North Central	pre-1978	1979	Merged within Southern to form Republic
		1986	Republic acquired by Northwest
Ozark	pre-1978	1986	Acquired by TWA
Piedmont	pre-1978	1987	Acquired by USAir
Southern	pre-1978	1979	Merged with North Central to form Republic
		1986	Republic acquired by Northwest
Texas International	pre-1978	1982	Acquired by Continental
USAir	pre-1978	1989	4th ranking carrier
Intra-state carriers (5)			
Alaska	pre-1978	1989	15th ranking carrier
AirCal	1979	1987	Acquired by American
Air Florida	1979	1984	Ceased operation due to bankruptcy
		1985	Acquired by Midway
PSA	1979	1987	Acquired by USAir
Southwest	1979	1989	9th ranking carrier
Charter carriers (2)			
Capitol	1979	1984	Ceased operation due to bankruptcy
World	1979	1985	Ceased operation due to bankruptcy
Commuter carriers (3)			
Air Wisconsin	1982	1989	18th ranking carrier
Empire	1980	1986	Acquired by Piedmont
Horizon	1983	1986	Acquired by Alaska
New carriers (17)			
Air Atlanta	1984	1986	Ceased operation due to bankruptcy
Air One	1983	1984	Ceased operation due to bankruptcy
American International	1982	1984	Ceased operation due to bankruptcy
America West	1983	1989	11th ranking carrier
Florida Express	1984	1988	Acquired by Braniff
Frontier Horizon	1984	1985	Ceased operation due to bankruptcy
Hawaii Express	1982	1983	Ceased operation due to bankruptcy
Jet America	1982	1986	Acquired by Alaska
Midway	1979	1989	16th ranking carrier
Muse (Transtar)	1981	1985	Acquired by Southwest

Table 2.1 Continued.

Origin and name	Began service*	Date	Status
New York Air	1980	1985	Acquired by Continental
Northeastern	1982	1984	Ceased operation due to bankruptcy
Pacific East	1982	1984	Ceased operation due to bankruptcy
Pacific Express	1982	1984	Ceased operation due to bankruptcy
People Express	1981	1986	Acquired by Continental
Presidential	1985	1987	Became feeder carrier for United
Sunworld	1983	1988	Ceased operation due to bankruptcy

* Date carrier began interstate service with jet aircraft.
** Size ranking based on passengers carried during 12 months ended September 1989.

proceeded extremely rapidly during the mid-1980s, it was primarily the result of forces that had accumulated during the years preceding and immediately following deregulation. In fact, the number of certificated trunk airlines had declined slowly throughout the history of CAB regulation, also as a result financial stresses on some pre-regulation carriers, from 16 in 1938 when the CAB was established to the 11 that operated during 1978. Similarly, the eight local service airlines in service at the time of deregulation represented slightly more than half of the 14 initially authorized by the CAB during the late 1940s. In retrospect, it may thus be reasonable to view the relatively short period (1981-84) during which the number of carriers proliferated, rather than the subsequent years of rapid consolidation in their number, as historically unique.

Perhaps the dominant factor contributing to this consolidation has been the previously-discussed economies inherent in network size, which appear to be quite substantial. In the openly competitive environment fostered by deregulation, some airlines have sought to realize these economies by merger or acquisition, rather than by internal growth, even at the risk of the transitional problems that can arise in integrating diverse operations. In some cases, it appears that their attempts to do so have compelled other carriers to adopt similar strategies, in order to avoid having their own opportunities for expansion pre-empted by their competitors' actions.

Another important factor promoting this development has been the successful efforts made by the pre-deregulation trunk carriers to reduce their operating expenses by controlling labour costs and increasing productivity. At the same time, these carriers have implemented a variety of competitive tactics – such as hub-and-spoke route networks and carefully controlled discount fare offerings – designed to enhance their revenues at the expense of competitors. Together these measures have not only made the major carriers less vulnerable to competition from the low-fare, no-frills competitors that proliferated during the early 1980s, but have actually reversed the competitive advantage briefly enjoyed by low-fare carriers.

Finally, the industry's recent consolidation has also come about partly in response to the uneven distribution of profitability among individual carriers (the history of which extends well into the regulated era). The cumulative financial distress of a number of carriers (particularly Eastern, People Express, Frontier, Muse, and Pan Am) became so great that they ceased to be viable competitors in most of the markets they served. Yet many of these financially troubled airlines held valuable assets – aircraft, takeoff and landing 'slots' at capacity-controlled airports, and long-term leases on airport facilities – that, together with their depressed values, made them attractive takeover targets for other airlines.

Development of hub-and-spoke route networks

The most often-noted development in the US airline industry during the decade of its deregulated operation has been the transition of individual carriers domestic route systems toward hub-and-spoke configurations. The nation's pre-deregulation route network, fashioned by the CAB through its awards of operating authority over 40 years, originally consisted of long-haul routes connecting major cities that were generally served by trunk carriers, together with localized networks connecting smaller cities to one another and to major cities in a region, which were operated by the local service airlines. Over time, some trunk and local service carriers were able to develop route hubs within the constraints of the Board's relatively restrictive route award policies. Although a few carriers experimented with different route development strategies in the period immediately following deregulation, within a few years each had begun to develop one or more major route hubs at large, strategically-situated airports.

The pre-deregulation route network

Over the four decades that it regulated US carriers' operating authority, the CAB appeared to exercise a consistent vision of how the nationwide air service network should be structured, and of the role that different carriers' route systems should serve. Routes connecting major cities, which were typically 500 miles or more in length and quite heavily travelled – primarily by passengers originating in and destined for the major cities comprising each route's end points – were considered to be the province of the large trunk carriers. The Board considered short-haul routes connecting cities within each region of the country to be a logical market for local service carriers, which it established experimentally in the late 1940s, subject to strict controls on their route authority that effectively limited them to providing such 'feeder' service (Eads, 1972).

Although the CAB's enthusiasm for the local service airline 'experiment' appeared to be limited, these carriers were granted permanent operating certificates by Congress in 1955. Subsequently, CAB adopted a policy of using route awards in an attempt to strengthen the local service airlines, although still without allowing them to compete directly with the trunk carriers. It attempted to do so selectively by transferring lightly travelled routes from trunk carriers to local service airlines, and by authorizing local service carriers to add service to previously unserved points in their geographic regions. When these initiatives failed to remedy the financial distress of the local service airlines, the CAB shifted to a policy of awarding non-stop service rights over longer routes to many of these carriers, while arranging for the weakest of them to be acquired by financially healthy carriers (Meyer & Oster, 1981).

In short, the Board's historical route award policies were motivated primarily by considerations of protecting what it saw as the 'natural' service areas of different types of carriers, and to a lesser extent by a desire to stabilize individual carrier's financial performance. Insofar as it was guided by a vision of a desirable route structure, the CAB sought to segregate long-haul from short-haul routes, as well as heavily patronized routes from lightly travelled markets, by authorizing different carriers to serve them. The pattern of route awards produced by these motivations apparently gave little consideration to allowing individual carriers to develop route networks that integrated short and long-haul service or routes of different traffic levels in an operationally rational manner.

Nevertheless, some carriers were succesful in developing limited hub-and-spoke route networks within the constraints imposed by the Board's route award policies. Despite the difficulties they faced in doing so imposed by the CAB's restrictive route award policies, 12 of the 15 trunk and local service airlines had each been able to develop at least one route hub where it operated 50 or more departures on an average weekday by 1976, and half of these operated 100 or more daily departures from at least one such hub (Meyer & Oster, 1981). These carriers' continuing efforts to solicit and obtain the collections of route authority necessary to develop major hubs – route award proceedings before the Board were often time-consuming and laborious, with repeated petitions sometimes necessary to gain approval to operate on an individual route – no doubt illustrated the operational value they saw an integrated hub-and-spoke network.

Accelerated development of hub-and-spoke networks

Under the terms of the Deregulation Act of 1978, any applicant found 'fit, willing, and able to properly perform air transportation' was required to be granted authority to operate on any route it sought to serve. In the months after its passage, the CAB processed carriers' applications for new route authority so rapidly that existing carriers were able to begin serving new routes without significant delays. At the same time, the process of withdrawing service from existing routes was dramatically accelerated by the Board's administration of the Act's provisions, so that carriers could restructure their route networks essentially without CAB interference.

In response to the flexibility granted by the Board's liberal interpretation of the 1978 Act's provisions, the pre-deregulation airlines experimented with a variety of route strategies. Although the most popular of these was the addition of new long-haul routes (which had been the most lucrative under the CAB's fare-setting policies), the most successful quickly proved to be the expansion of existing hub operations undertaken by some smaller trunk and local service carriers. These carriers expanded their pre-deregulation hub operations primarily by adding service from those hubs to cities outside their traditional service areas, while reorganizing scheduled arrivals and departures at these hubs to facilitate transfers between flights by passengers who were not destined for the hub city.

By the early 1980s, virtually all of the pre-deregulation carriers and new entrants into the industry had begun to aggressively expand their operations to and from a major hub city. In some cases, this process involved developing a major route hub at a city that had been a relatively minor point in a carrier's pre-deregulation network. A few carriers who operated large hubs prior to deregulation not only expanded these rapidly, but also developed major hubs at a second city. Thus by 1984, 13 of the 15 pre-deregulation trunk and local service carriers operated at least one hub with 100 or more daily departures, with seven of these accounting for 200 or more departures. In turn, four of these seven carriers also operated 100 or more daily departures at a second hub by 1984 (Meyer & Oster, 1987).

Since 1984, continued hub development and mergers have combined to consolidate the US domestic air service network around a set of 'mega-hubs', while the number and importance of carriers' secondary hubs have also continued to grow. Thus by 1988, nine of the 10 remaining large airlines operated 200 or more departures at a single hub, while an additional three operated two hubs of this size, and six of the 10 also operated 100 or more daily departures at one or more secondary hubs. The nation's system of hub airports had also become extremely 'Balkanized' by this time, with each of the four largest hub cities (Atlanta, Chicago, Dallas, and Denver) serving as a major hub for two carriers, while operations at each of the nation's remaining major hubs were dominated by a single large carrier.

Integration of local service into hub operations
In addition to operating short and long-haul jet service to cities spanning a wide range of sizes from its major hub (or hubs), each carrier has also attempted to integrate into its route network, service to small communities in the immediately surrounding area that are too small to support acceptably frequent flights using its own jet aircraft. This service was historically provided by commuter (now generally called regional) airlines, using propeller-driven aircraft in the 10–60 seat range. Through joint marketing agreements with regional airlines operating in the areas surrounding their route hubs, major carriers have sought to incorporate connecting flights to these small communities into the networks of service they can offer through their hub cities.

One controversial element of these joint marketing agreements is 'code-sharing', whereby flights operated by regional airlines are identified in the computer reservation systems used by travel agents (as well as in the printed

Official Airline Guide) under the same codes used to identify the flights of their major airline marketing partners. Code-sharing became controversial because it initially led some travellers to believe erroneously that service to and from small communities would be operated by the major carrier under whose code it was advertised. In addition, the degree to which major and regional carriers who participated in code-sharing relationships actually co-ordinated their flight schedules, through ticketing arrangements, and passenger and baggage handling activities varied widely. More recently, however, this controversy has subsided, probably in response to clearer identification of the actual operator of such flights in computer systems and printed schedules, coupled with closer co-ordination of operations between code-sharing major and regional airlines.

Marketing airline services through computer-equipped travel agents

The third major industry-wide development in the era of deregulated airline competition has been a widespread shift away from direct marketing and distribution of air travel service by airlines themselves towards travel agents equipped with computerized passenger reservation and ticketing systems, who serve as distribution channels for virtually all airlines. Like the development of hub-and-spoke networks, the shift from direct airline marketing to the use of travel agents began during the regulated era, but accelerated with the industry's deregulation. At the same time, computer reservation systems (CRSs) – first made available to travel agents in the mid-1970s – have grown to become a virtually universal feature of the distribution system for air (and other) travel services.

The increasing importance of travel agents
Even in the regulated era, airlines relied partly on travel agents to book passenger reservations and issue tickets, since agents provided a ubiquitous, relatively efficient and low-cost marketing alternative to their own internal reservations agents. Thus by 1976, the airline industry was already served by some 12,000 travel agency locations (travel agencies often have multiple locations, each of which employs one or more individual agents) throughout the nation (US General Accounting Office, 1986). With the proliferation of the number and range of fares spawned by deregulation, however, travel agents have become of increasing potential value to travellers, while competitive pressures for cost control have made agents an even more attractive marketing alternative from the viewpoint of the airlines themselves.

Thus it is not surprising that the number of travel agents, as well as the airline industry's and travellers' reliance on them as a distribution channel, have both increased dramatically in the decade following deregulation. By 1986, the number of agency locations nationwide had reached at least 25,000, more than double the 1976 figure (some estimates indicate that there may have been as many as 30–35,000). During that same period, the fraction of the total dollar value of airline tickets sold through travel agents grew from approximately 40% to nearly 85%.

Growing reliance on CRSs

The flexibility to add service to new cities without a lengthy approval process, together with the increasing variety of fares and service classes offered by airlines in response to deregulation has also made it virtually essential for travel agents to have instantaneous access to airlines' schedules and fares. Computerized reservation systems capable of providing this access were developed by several large airlines starting in the mid-1970s, as outgrowths of these carriers' earlier investments in computerizing their internal reservation systems. United and American Airlines first provided their CRSs to travel agents (who lease equipment and software and obtain instruction in their use from their airline vendors) in 1976, and within a few years five reservation systems were being commercially vended throughout the US by their airline owners.

Because of the increasing role they played in marketing air travel, airlines quickly realized the importance of providing travel agents with the ability to book reservations and sell tickets for travel on their flights using whichever of these systems agents chose to employ. Since even the smallest of the five systems is subscribed to by substantial numbers of travel agents, virtually all airlines – including those that own their own CRSs – now provide travel agents with access to their flights through all five systems, in exchange for which they agree to pay the owner of each system a fee for every passenger reservation made by a travel agent using it. The technological capabilities of these systems have continued to evolve rapidly, so that each system can now display to travel agents an up-to-date inventory of available seats by fare class on every flight operated by airlines participating in the system, as well as automatically issue both tickets and boarding passes, in addition to accepting and recording passenger reservations.

In combination with the expanding role of travel agents, the increasing importance of CRSs to both airlines and agents has caused the use of these systems to expand dramatically. During 1980, approximately 7,500 travel agent locations – roughly one-quarter of the total number that existed – were equipped with one of the three airline-owned CRSs then available. Sales by agents through these three systems accounted for less than 5% of the US airline industry's total passenger revenues during 1980. Yet by 1986, nearly 25,000 travel agency locations, representing more than 95% of all those nationwide, were equipped with at least one of the five airline-owned CRSs. During that year, agents used these systems to book more than half of the airline industry's total dollar value of ticket sales.

For the foreseeable future, it appears likely that CRS-equipped travel agents will continue to play a critical role in marketing and distributing air travel. Although currently available technology (principally automatic ticketing machines that could be used directly by travellers) could facilitate major changes in the marketing and distribution of airline services, airlines are reluctant to promote its adoption for two reasons. First, such technology offers the potential to bypass travel agents, whom airlines seek to accommodate, and second, the complexity of fare and service offerings – which has distinct attractions for airlines, but could require time-consuming efforts by travellers to search and choose among – may deter widespread use of direct sales outlets.

2.4 The effects of deregulation on fares and service

The major industry-wide developments spawned by airline deregulation have in turn given rise to pronounced changes in the overall level and structure of fares for air travel, as well as in the pattern and quality of airline service. From the viewpoint of travellers, these are of course the most visible consequences of the decade-old change in policy, and are thus the primary criteria on which the general public – and, not incidentally, its elected representatives – have and will continue to judge its success or failure. From the social scientist's perspective, changes in fares and service in response to deregulation are also important, since they contribute by far the largest share of the change in social welfare stemming from the change in policy.

Unfortunately, assessing the impacts of deregulation on airline fares and service is not simply a matter of measuring the changes that have occurred since the policy began (and even determining that date can be complicated). Because fare levels and service patterns were decidedly not static during the era of CAB regulation, it is difficult to imagine how they would have changed if the regulatory regime had persisted through the present, and thus to establish a 'counterfactual' case against which to measure deregulation's impact (Morrison & Winston, 1986).[12] Nevertheless, the 40-year record of regulation as practised by the Board does provide useful indications about the course of hypothetical fares and service under continuing regulation, which can be used to develop rough estimates of the effects of its demise.

Cost growth and airline fare increases

Prior to its approval during 1977 of several carriers' applications to offer widely available discount fares, the CAB had historically adjusted fares upward in response to increases in the airlines' average unit expenses for operating their services. In order to do so, it monitored carriers' average operating expenses, and usually authorized across-the-board increases in fares when the industry-wide level of these expenses rose.[13] Recognizing that airlines' unit operating expenses decline with increasing flight length, the CAB authorized per-mile fares that also declined as market distance rose, although it did not permit fares to decline with distance as rapidly as did costs.[14] Other than allowing lower per-mile fares on longer routes, however, regulators permitted little variation among fares in different markets, even where their characteristics (such as the number of passengers travelling on a route) introduced differences in the costs of serving them (Kaplan, 1986).

The CAB's fare-setting deliberations also tended to focus primarily on setting prices for standard coach-class airline service. Although it occasionally permitted carriers to offer limited discounts from full coach fares, the Board generally set detailed terms governing passengers' eligibility to travel at reduced fares. In addition, regulators generally stated the discounts they permitted as a percentage reduction from the coach fare. Thus both the entire structure of fares and the average fare actually paid by all passengers, including those travelling at full coach and discount fares, tended to vary in response to the changes the Board authorized in coach fares.

Airline productivity before and since deregulation
Table 2.2 compares pre- and post-deregulation increases in an index of prices paid for the inputs carriers utilize in producing airline service, their average expenditures per seat-mile, and the average fare paid per passenger-mile of air travel. As it indicates, both carriers' input prices and unit operating expenditures increased rapidly in the years preceding de-regulation, and have continued to do so since then.[15] Nevertheless, the airline industry has been successful in offsetting the effects of rising input prices by increasing the productivity with which it employs those inputs in operating its services, both before and since deregulation. However, the rate at which higher unit operating expenses have been translated into higher overall airfares appears to have slowed since deregulation began.

The table shows that during the last decade before the CAB began to deregulate fares (1967–77), airlines' input prices escalated 111%, yet only about half of this increase was translated into higher costs per seat-mile, which rose 55% during the same period. Improvements in the productivity with which airlines operated must have proceeded rapidly over this period to account for

Table 2.2 Changes in airline input prices, operating expenses, and average fares.

Year	Airline input price index (1967 = 100)	Expense per seat-mile (cents)	Average fare per passenger-mile (cents)
1967	100.0	3.10	5.49
1977	210.9	4.79	8.42
1987	388.5	7.53	11.10
Change:			
1967–77	111%	55%	53%
1977–87	84%	57%	32%

Sources: Air Transport Association. *The Annual Report of the US Scheduled Airline Industry* (various years); Airline Economics, Inc. *The Airline Quarterly* (various issues).

this pronounced difference between the rate of input cost increases and growth in unit operating expenses. This was no doubt partly the result of continued conversion of the industry's aircraft fleet to jets, which comprised 59% of the total during 1967 but nearly the entire fleet by 1977, coupled with the introduction of wide-body jet aircraft in 1968, which represented nearly 15% of the US airline fleet by 1977.[16] Because unit operating expenses also decline with flight distance, which increased nearly one-third over this period, operating longer flights may also have been a source of productivity growth.

Table 2.2 also shows that during the first decade of deregulated fare competition (1977–87), airlines have continued to absorb rapid input price increases through productivity improvements. Approximately two-thirds of the 84% increase in airline input prices over this period shown in the table was reflected in higher unit operating expenses, which rose 57%. Thus the industry's productivity in operating airline services must have continued to improve since deregulation began, although apparently at a slower pace than during the previous decade. The post-deregulation record of continuing productivity improvements is nevertheless impressive, because the rapid pace of technological innovation during the latter years of the regulatory regime slowed dramatically during this period, while increases in the average length of flights were also a less important source of downward pressure on airlines' unit costs.[17] Much of the continuing growth in productivity must therefore owe to incentives established by the price competition inaugurated by the CAB and later codified in the 1978 Deregulation Act.

The effect of deregulation on fares

The number of fare classes has grown dramatically since deregulation, and most travel now occurs in these new classes. As a result, it is nearly impossible to construct a conventional price index that measures the overall change in fare levels. In the absence of such an index, most analyses of the effects of deregulation focus on the average fare actually paid by travellers, as measured by average fare revenue per passenger-mile travelled. Variation in this measure captures the effects of both changes in the levels of fares for different classes of travel, and changes in the numbers of passengers travelling in the different fare classes. While it is thus an imperfect measure of changes in fares charged for airline of a constant quality, the average fare is still a useful index of movements in the overall price level of air transportation.

Table 2.2 shows that during the last decade of regulation the CAB's fare-setting policies resulted in increases in the average fare paid by air travellers totaling 53% between 1967 and 1977, thus almost exactly matching those in airlines' unit operating expenses. Of course, the direction of causality in this relationship is difficult to isolate, since the Board's practice of authorizing fare increases to compensate for escalating operating expenses may have blunted carriers' incentives to control their expenses. Regardless of the specific causes of rising costs, regulators clearly permitted carriers to raise fares for air travel almost exactly in tandem with growth in their operating expenses.

Since the Board's initial steps to deregulate fares in 1977, however, this pattern has changed distinctly. Table 2.2 reports that while industry-wide unit cost increases during the 1977–87 period were slightly higher than during the preceding decade (57% versus 55%), the average fare charged by airlines rose by only 32%. In contrast to the regulated period, when fare increases closely paralleling unit cost growth were apparently authorized, only somewhat more than half of the post-deregulation increase in carriers' unit costs has been passed through to travellers in the form of higher fares. Thus the competitive pressures unleashed by deregulation have apparently been effective in reducing the proportion of exogenous increases in airlines' input prices that is ultimately translated into higher fares for travellers.

It seems reasonable to presume that if CAB regulation of the industry had continued, the increase in carriers' unit operating expenses between 1977 and 1987 would have been fully reflected in higher fares, as was the case in the prior decade. Under this assumption, the average fare actually paid by US air travellers during 1987 would have been some 57% (equal to the 1977–87 increase in carriers' unit operating expenses) higher than its 1977 level, or about 13.2 cents per mile. The actual 1987 average fare level of 11.1 cents shown in Table 2.2 is roughly 15% lower than this hypothetical 1987 regulated fare level. Although this figure should not be construed too literally, it does provide a rough estimate of the impact deregulation has exercised on the setting of airfares.

Further, if any part of the continuing post-deregulation improvement in airline productivity is attributed to the more competitive environment in which carriers now operate, it is plausible that regulated fares during 1987 would have been even higher than this estimate, and thus that deregulation has reduced actual fares by more than the 15% figure. On the other hand, fare increases have slightly outpaced both input price increases and growth in carriers' unit operating expenses since 1987, thereby eroding the gap between actual and hypothetical regulated fares. Thus on balance, it seems reasonable to believe that the overall level of airfares today is approximately 15% lower than would be the case under continued CAB fare regulation conducted according to the policies it employed during its last decade of actual authority.

Other estimates of deregulation's effect on fares
Although this estimate of fare savings from deregulation is slightly lower than previous estimates, it agrees surprisingly closely with more recent analyses of deregulation's effect on the level of airfares. Call and Keeler (1985) estimated that for the 100 most heavily travelled domestic routes, the lowest available fare without advance purchase or other restrictions averaged 15% below what it would have been with continued regulation during 1980, and 20% below the regulated level during 1981. These authors calculated post-deregulation values of the index used by the CAB in setting fares during the immediate pre-deregulation period, and compared changes in this index to fare changes that were actually observed in a sample of markets. Using a deflator of their own

design to adjust for the effect of increases in carriers' input price increases following deregulation, Morrison and Winston (1986) estimated that deregulated airfares during 1977 would have been about 28% below their actual level during that year.

More recently, both the US Department of Transportation and Morrison and Winston (1989) have developed separate estimates of likely trends in the CAB's fare index since deregulation began in 1977. The former employs the actual 1977–88 rate of productivity growth among all US airlines – including carriers that entered inter-state service during the deregulated era – to calculate recent values of the index. Its results suggest that average actual fares have varied from 5% to perhaps as much as 20% below hypothetical regulated fares during the 1980s, and were approximately 10% below those that would have resulted from continuing regulation during 1988. Morrison and Winston assume that the CAB would have continued to prohibit entry into the industry by new carriers, and that productivity under continued regulation would have grown at the slower actual rate exhibited by regulated foreign airlines. Under these assumptions, they estimate that actual fares during the 1977–86 period averaged 18% below those that would have prevailed under continuing CAB regulation, but have stabilized at approximately 13–15% below hypothetical regulated fares in recent years.

Thus it appears that the figure of 15% savings does represent a reliable estimate of the current effect of deregulation on the overall level of airline fares. However, it also appears that the difference between actual fare levels and those under continuing CAB regulation may have peaked during the early 1980s, when actual fares were by several estimates as much as 20–25% below hypothetical regulated levels. After declining to the 15% neighbourhood by the middle 1980s, the difference between actual and hypothetical regulated fares appears to have remained quite stable, with average fares varying primarily in response to changes in carriers' input costs, primarily those for labour, capital financing, and fuel.

Deregulation and the structure of airfares

Certainly the most pronounced effect of deregulation on prices for air travel has been a dramatic increase in the number and range of fares that are available for travel on most routes. The wide range of fares now available tends to be distinguished by differences in the conditions attached by airlines to their use, such as advanced purchase requirements, together with restrictions on the ability to change travel schedules and itineraries (increasingly including cash penalties). In addition, the decline in per-mile airline fares with distance travelled — intentionally suppressed by CAB regulation in its desire to cross-subsidize short-haul service with profits from long-haul routes - become more pronounced in the deregulated era.

Variation in fares about their average level

One consequence of the proliferation of fare and service classes since deregulation has been a pronounced expansion in the range of fares that are utilized by passengers travelling in most markets. Table 2.3 compares the distributions of actual airfares about their average values during 1978 and 1988, which provides one indication of how the structure of fares has changed in the wake of deregulation. As it indicates, nearly two-thirds of passengers paid fares within 20% (above or below) of the average fare charged on the routes they travelled during 1978, yet by 1988 this fraction had declined to slightly more than one-third (35%). While only about 20% of passengers paid fares between 40% and 80% of the average on the routes they travelled during 1978, this fraction had nearly doubled by 1988. Yet as the Table also shows, the fraction of air travellers paying fares equally far above their average level for comparable trips (that is, from 120 to 160% higher than average) increased only slightly over this same period. Finally, Table 2.3 shows that the fractions of passengers paying both extremely high and extremely low fares rose over the decade following deregulation, with nearly 10% paying fares 60% or more above the average paid on the routes they travelled by 1988, a category that probably includes most first class and unrestricted coach-class fares.[18]

Table 2.3   Distribution of fare levels during 1988.

Percentage of average fare	Passengers in fare category	
	during 1978 (%)	during 1988 (%)
under 40%	0.2	1.5
41–80%	19.7	38.9
81–120%	66.2	35.0
121–160%	12.5	14.8
above 160%	1.4	9.8

Source: Calculated from information provided by Steven Morrison in private communication to the author, 19 January 1990.

Thus Table 2.3 shows that the current distribution of fares appears to be considerably 'flatter' than during the regulated era. The primary cause of this flattening is the proliferation of discounted fares that airlines have used the pricing freedom permitted by deregulation to make available, which is reflected in the pronounced shift from the 81–120% of average category shown in the Table to the 41–80% interval. A substantial fraction of airline passengers is apparently willing to accept the advance purchase, fixed travel schedule, and other restrictions – increasingly including cash penalties for cancellation or schedule changes – accompanying these discounts in exchange for what appear to be considerably lower fares than would have been available under continued regulation. In contrast, fares for premium service (first class) and unrestricted travel appear to be substantially higher than was the case under CAB regulation.[19]

Variation of fares with distance travelled

Table 2.4 illustrates the current variation in airfares with distance travelled, and compares this to the pattern of fares that would have resulted from continued adherence to the CAB's fare-setting formula. As the Table indicates, actual per-mile fares declined sharply with distance travelled during 1988. On routes ranging from 750–1,000 miles, fares averaged well under half of their level on routes under 250 miles, and on trips over 2,000 miles, declined to less than one-sixth of their level in the shortest route category.[20] While fares set according to the CAB formula (adjusted to reflect carriers' 1988 operating expenses) would also have declined with distance, they would have fallen by only about 50% from the shortest (under 250 miles) to the longest (over 2,000 miles) mileage intervals shown in the table. Thus actual fares during 1988 were as much as one-third higher than hypothetical regulated fares would have been on very short routes, but were almost 50% below regulated levels on the longest trips, as Table 2.4 shows.

Table 2.4 Variation in actual and hypothetical regulated fares by distance interval during 1988.

Distance interval (miles)	Average actual fare ($ per mile)	Hypothetical regulated fare ($ per mile)	Actual as a percentage of regulated fare (%)
under 250	0.37	0.28	132
250–500	0.26	0.22	118
501–750	0.21	0.18	117
751–1,000	0.16	0.17	94
1,001–1,500	0.13	0.15	87
1,501–2,000	0.09	0.14	64
over 2,000	0.07	0.13	54

Source: Calculated from information reported by airlines to US Department of Transportation.

Although the resulting pattern of fares over distance during 1988 should provide a more accurate reflection of how airlines' costs vary with distance, there is some concern that in direct contrast to the regulated era, current fares may now be above costs for short trips and below costs on longer routes. This concern arises because the average number of competitors was higher on longer routes during 1988 than in short-haul markets, while within each distance category fares generally decline as the number of competitors increases. Yet actual 1988 fares were still above their hypothetical regulated levels on most competitive routes under 500 miles, and even in monopoly markets, fares were still well below costs at distances above 1,000 miles. Hence it does seem safe to regard the current pattern of US airfares over distance as a reasonably accurate reflection of how costs differ by route length, and thus of how fares might ultimately be expected to vary in other contexts where deregulation is contemplated.

Changes in service levels since deregulation

Unfortunately, it is even more difficult to anticipate how the current pattern of airline service under continued regulation would differ from that actually observed than was the case with airfares. The CAB evaluated carriers' petitions for new operating authority on an individual, route-by-route basis, and once authorized to serve a route, carriers were free to vary departure frequencies at their own discretion. As discussed previously, during the early 1970s the Board had taken important steps to stabilize the pattern of service then prevailing: it unofficially established a moratorium on new route authority, and granted anti-trust immunity to the largest carriers in order to encourage them to discuss mutual capacity reductions on a number of routes (Kaplan, 1986). Thus a reasonable assumption seems to be that the current pattern of service under continued regulation would resemble that prevailing when the Deregulation Act was passed (in October 1978), and that changes in the level and pattern of airline service since that time reflect primarily the influence of deregulation.

Overall changes in service frequency
Table 2.5 illustrates changes in the frequency of aircraft departures from the 183 US airports that received scheduled service during July of 1979 from carriers regulated by the CAB.[21] As it shows, the frequency of service to 137 (or almost exactly three-quarters) of these airports increased between 1979 and 1988, in many cases by 50% or more. At the remaining one-quarter of these 183 points, declines in the frequency of service – most commonly in the 10–50% range – occurred over this period. By comparison, service frequency increased at less than one-third of the 445 other airports that had been served by *un*regulated

Table 2.5 Changes in frequency of airline service 1979–88.

	Number of points	Percent of points
Service increases		
under 10% *	18	9.8
10–50%	59	32.3
50–100%	43	23.5
over 100%	17	9.3
Subtotal	137	74.9
Service reductions		
under 10%	9	4.9
10–50%	30	16.4
over 50%	7	3.8
Subtotal	46	25.1
Total	183	100.0

* Includes points with no change in frequency of service.
Source: Official Airline Guide (various issues).

carriers during 1979, although this comparison must be interpreted cautiously since the characteristics of travel and service differ between these and formerly regulated markets.

Service changes at cities of differing sizes

Table 2.6 shows that the frequency of flight departures increased considerably between 1977 and 1984 at cities in every size category, rising by more than 43% at the largest points and by 20% even at airports serving the nation's smallest communities.[22] Thus in total at least, the reductions in service to these small communities foreseen by some opponents of deregulation certainly did not materialize during the early years after it began. The table also reports that while the average seating capacity of flights departing from the nation's largest cities increased during the first several years of deregulation, it declined at cities in every other size category, falling substantially (by 22.7%) in the smallest city size category.

Table 2.6 Changes in service frequency and aircraft size by city size category, 1977–84.

City size category	Percentage change in flight departures	Percentage change in seats per departure
Large	43.4	2.3
Medium	41.0	− 5.7
Small	31.6	− 10.2
Other*	20.0	− 22.7
All city sizes	36.3	− 0.4

* Includes non-urban areas and cities under approximately 100,000 in population.
Source: Computed from information reported in US Congressional Budget Office (1988).

Some of the increased service at communities in the two smallest size categories included in Table 2.6 thus apparently represents the substitution of more frequent flights using the smaller, propeller-driven aircraft operated by unregulated airlines for the often infrequent jet services these communities received from regulated carriers during the pre-deregulation era. Oster and Zorn (1987) report that in 60 city-pair markets where this occurred, average weekday departures rose from fewer than three to well over six. (Even before deregulation, however, much of the service to these points was already provided by these unregulated 'commuter' airlines, and the CAB had begun to allow the carriers it regulated to withdraw service from many smaller points.) Although the overall effect on the quality of air service received by communities where deregulated allowed this substitution to accelerate is a subject of continuing controversy, it is nevertheless clear that one – and perhaps the most – important dimension of the convenience offered by airline service to these communities has improved, often significantly.

Service changes by market size

Evidence of the effect of deregulation on service in individual market categories is more difficult to compile and to interpret, primarily because the accelerated development of hub-and-spoke route networks discussed previously has resulted in a considerable increase in connecting service, which is difficult to enumerate from published or computerized flight schedules. Table 2.7 compares two previously reported (Morrison & Winston, 1986; US General Accounting Office, 1986) measures of service changes in markets connecting cities of different sizes during the first several years of deregulation, one of which includes and the other excludes connecting flights. It shows that the frequency of direct service (non-stop flights plus those entailing stops but not changes of aircraft) increased sharply through 1984 in markets connecting medium and large cities with one another and with smaller points. Direct service connecting these smaller points with one another, however, generally declined in the years following deregulation, although this decline was substantial in only one size category (small to non-hub markets).

Table 2.7   Changes in service frequency by market size category.

Market size category	1977–84 change in direct flights * (%)	1977–83 change in direct plus connecting flights (%)
Large–large city	17.8	− 3.5
Large–medium city	22.1	14.4
Large–small city	22.2	19.2
Large city–other**	14.7	28.7
Medium–medium city	33.5	− 4.3
Medium–small city	25.5	20.8
Medium city–other	1.2	24.3
Small–small city	− 2.9	33.9
Small city–other	− 16.9	1.4
Other–other	− 6.9	33.9

   *   Includes non-stop flights plus flights making stops but requiring no change of aircraft.
   ** Includes non-urban areas and cities approximately 100,000 in population.
   Source: US Congressional Budget Office (1988); Morrison & Winston (1986).

When connecting flights are included, however, Table 2.7 shows that there were increases in the frequency of service in virtually all market size categories.[23] In particular, increases in connecting service appear to have been more than sufficient to compensate for losses in direct flights in markets connecting small and non-hubs with each other. In addition, differences in the two measures of service changes in large hub to non-hub and in medium hub to non-hub markets suggest that the frequency of connecting flight opportunities also increased sharply in these two market categories. Much of the growing frequency of connecting service to small and non-hub communities is probably accounted for by carriers' reorganization of their route systems around major hubs, which substitutes more frequent service via these hubs for the typically less frequent direct service between them that resulted from common aircraft routings over the more linear route networks of the regulated era.

It is difficult to determine how the pattern of service frequencies in different markets has evolved since the dates (1983 and 1984) covered by the measures reported in Table 2.7. Between 1984 and 1988, aggregate flight departures increased another 23% (Air Transport Association, 1989), bringing their overall increase during the deregulated era to more than 67%. However, some of the post-1984 increase in flights is no doubt accounted for by carriers' efforts to accommodate the growth in travel demand spurred by the US economy's emergence from its prolonged recession of the early 1980s, which cannot of course be credited to deregulation. Further, certain mergers between large carriers operating hubs at the same large hub cities that occurred during 1985 and 1986 resulted in service reductions on many routes to and from those hubs, but were also accompanied by increases in service connecting these large hubs with surrounding small communities (US General Accounting Office, 1986). Thus on balance, the changes in overall service levels and the pattern of these changes by market category shown in Tables 2.5 and 2.6 may portray reasonably accurately the changes in airline service owing to deregulation *per se*, as distinct from those caused by other developments not associated with the change in regulatory policy.

Deregulation and airline safety

The safety record established by US air carriers that were historically subject to regulation of fares and service has continued to improve since the CAB took its first steps to relax control over fares and service, as it did throughout the four decades of regulation by the Board.[24] The rate at which fatal aircraft accidents occurred in scheduled domestic service declined by more than 50% (from 0.46 to 0.22 per million departures) between the years preceding deregulation and the period since passage of the 1978 Act, while the rate of occurrence of accidents causing serious injuries declined even more rapidly (Oster & Zorn, 1987). In fact, analysis of yearly variation in accident rates suggests that the pace of improvement in airline safety has probably accelerated since economic deregulation of the industry (Rose, 1987). Other measures of safety performance, such as the number of fatalities and serious injuries per passenger trip, also showed similar declines between the pre- and post-deregulation periods.

Because service operated with smaller, propeller-driven aircraft is somewhat less safe than that provided by large jet-equipped airlines, passenger safety may have declined in markets where such 'commuter' airline service replaced jet service that major carriers were allowed by deregulation to withdraw. At the same time, however, the lower fares resulting from deregulation have caused a significant amount of intercity travel that would otherwise have used private autos or other common carrier modes to instead be made by air (Morrison &

Winston, 1986; Meyer & Oster, 1987). Because airline travel is by far the safest of these modes, its deregulation may thus have contributed indirectly to an increase in the overall safety of intercity passenger travel throughout the nation. In any case, it certainly does not appear, as some critics of the policy have argued, that economic deregulation of major airlines has caused a decline in the safety level afforded by the nation's air transportation system.

2.5 Problems and prospects

Viewed from the perspective of slightly more than a decade's experience, the US experience with deregulation of its domestic air transportation system appears on balance to have been quite successful. Although critics of the policy have seized upon the industry's rapid consolidation during the mid-1980s as evidence of an impending failure of deregulation, the number of airlines now offering service over comprehensive, nationwide route networks is at least equal to the number doing so prior to deregulation. More important, these carriers now compete with one another in more markets than ever before, offering service that is more comprehensive as well as generally more frequent. Further, the US industry's post-deregulation record of productivity improvement has nearly matched that enabled by the remarkable succession of technological innovations that occurred during the regulated era, resulting in significantly lower fares than would prevail today under continuing regulation. And despite claims to the contrary, the industry's safety record has improved at least as rapidly as under regulation during the process of its adaptation to the more competitive environment unleashed by deregulation.

Nevertheless, there remain several important concerns regarding the prospects for the policy's continued success. Perhaps most important, airlines' development of large hub-and-spoke route networks, in conjunction with marketing devices such as frequent-flyer programmes, may already have erected important barriers to entry by new competitors into both the airline industry itself and individual cities or city-pair markets. In addition, while airline-owned computer reservation systems provide important benefits to travel agents and airlines, they appear to impede competition among airlines in various ways. Finally, difficulties faced by potential competitors in gaining access to airport facilities on terms comparable to those enjoyed by incumbent airlines may place them at a critical disadvantage or preclude their entry entirely. Since the continued success of deregulation in maintaining lower fares and sustaining high-quality airline service depends critically on maintaining vigorous competition among airlines, the seriousness of these apparent impediments is of vital concern. At the same time, however, the potential effectiveness of available measures in mitigating these barriers, as well as any sacrifice in benefits likely to result from restricting the practices that create them, warrants careful consideration in designing remedial actions.

Hubs and marketing devices as entry barriers

Certain tactics adopted by airlines seeking to enhance their competitiveness in the deregulated environment have proved so successful that their resulting prevalence threatens the continued vigour of the rivalry unleashed by deregulation. These include the development of large route hubs at strategically-situated airports, frequent-flyer programmes, and exclusive joint marketing agreements between major carriers and smaller regional airlines. It is important to anticipate the potential longer-term effects of such strategies on the continued viability of the more intense competition spawned by deregulation, and to identify measures that can restrict these tactics' effectiveness in limiting the competitive opportunities remaining available to airlines seeking to begin or expand service. Because each of them has also produced substantial benefits for travellers, however, it is equally important that these be weighed against the benefits from potentially intensified competition in deciding whether such measures should be adopted.

Hub-and spoke route networks
The large hub-and-spoke route networks that have come to dominate the pattern of US domestic airline service during the deregulated era confer both costs and marketing advantages upon airlines that operate them. Such networks allow airlines to realize economies in operating both aircraft and airport facilities, while attracting higher passenger volumes with frequent service in a large number of markets. By doing so, however, these networks also raise the geographic scope and frequency of service with which a competitor would need to enter the airline industry, or inaugurate service at a city where another already operates a hub, and offer viable competition.

Thus both the total investment necessary to inaugurate this minimum scope and level of service and the fraction of that investment that is at risk (and could not be recovered in the event the entrant decided to withdraw or redeploy its services) are likely to increase in rough proportion to the scale of incumbent services against which it seeks to compete. New competitors operating below this critical threshold of scale are likely to be disadvantaged relative to larger incumbents by some combination of higher operating costs and lower passenger loads, except perhaps in providing service to new cities from an existing hub.

Frequent-flyer programmes
In combination with the development of large hub-and-spoke networks, frequent-flyer programmes – which provide free travel on an airline as a reward for accumulating extensive paid travel on its flights – also appear to reduce the threat of potential competition. Most of these programmes incorporate features that are carefully designed to encourage travellers to concentrate their travel on a single airline, such as increasing rewards for the accumulation of mileage increments and limits on the time period during which bonuses can be earned and used. Hence both the ease with which travellers can earn such

bonuses and their attractiveness once earned are greatest for the airline operating the largest proportion of service at a city, and particularly for one that operates a major route hub there. Widespread use of these programmes – those of the eight major US carriers now total some 34 million members – can thus handicap smaller carriers' efforts to compete in those markets, while making it more difficult for potential competitors to begin providing service there. Meanwhile, the free travel provided to participants in these programmes is paid for by higher fares on other travel, as well as perhaps in the form of 'excessive' flying engaged in by those who accumulate bonuses in the course of travelling at the expense of their employers (Stephenson & Fox, 1987; Tretheway, 1989).

At the same time as the popularity of these programmes inhibits entry into new markets by potential competitors, it is likely to make efforts to restrict their use or terms extremely unpopular. The most often-advanced proposal is to subject the value of free travel earned through frequent-flyer programmes to taxation as income, under the rationale that they are primarily earned in the course of employer-paid business travel and thus represent in-kind employee compensation of the sort normally subject to income taxation (Levine, 1987; Borenstein, 1988; Morrison & Winston, 1989). Yet bonuses earned as rewards for travel paid for by passengers themselves (many of whom who also travel at their employers' expense) should be exempted from taxation under this rationale, and separately valuing the two components would be particularly difficult under the graduated bonus schedules offered by most airlines' programmes. A more promising avenue, but one that would be predictably even less popular among airlines offering frequent-flyer programmes, would be to require that credits toward free travel be transferable among passengers. Although this would not completely eliminate the marketing advantage stemming from a large route network, it could substantially reduce the competitive advantage enjoyed by established carriers in the cities where they operate large hubs, thereby reducing the effectiveness of the programmes in deterring potential entry by new competitors.

Joint marketing agreements and code-sharing
Joint marketing agreements between regional and major airlines serving the same city (discussed previously in section 2.4) almost universally entail code-sharing between the two carriers, while also limiting the ability of the regional carrier to enter into similar agreements with other major carriers serving the same city. Both of these features may make it more difficult for other airlines to begin or expand service there, thereby reducing the effectiveness of potential competition in constraining fares and ensuring adequate service levels. Because code-sharing causes itineraries involving connections between flights operated by carriers that engage in it to be listed more prominently in CRSs than equally convenient connections between other airlines' flights, a major carrier without such agreements with local regional airlines is less likely to pose a credible competitive threat to an incumbent airline with such agreements already in

place. This potential threat is further reduced by the fact that virtually all US regional airlines now participate in such exclusive joint marketing and code-sharing agreements with a major airline at each large city they serve, making it more difficult for potential entrants or carriers seeking to expand service to attract the important 'feed' traffic provided by those regional airlines.

At the same time, however, these agreements provide some important benefits to passengers travelling to and from the small communities typically served by regional airlines. Regional and major airlines participating in them typically co-ordinate their flight schedules to minimize delays experienced by connecting passengers, locate their boarding gates proximately in airports (sometimes even sharing gates), and simplify through-ticketing and baggage checking for passengers connecting between their flights. Thus any action to restrict the scope or terms of these agreements should carefully weigh any reduction in these benefits that might result against the potential for increased competition. One reasonable course would be to prohibit the imposition of arbitrary penalties on connections between flights operated by different carriers in constructing and ranking travel itineraries for display on CRS screens. This prohibition would eliminate the attractiveness of code-sharing without reducing the benefits provided to travellers by other features of joint marketing agreements, although it would also increase the effort required for travel agents to search through CRS displays to identify connections between flights operated by carriers offering those features.[25]

Computer reservation systems and airline competition

Passenger diversion to CRS-owning airlines

The growing importance of travel agents equipped with airline-owned CRSs in marketing and distributing airlines' services poses two critical impediments to vigorous competition among airlines. Most important, travel agents continue to book more passengers for travel on the airlines that own the systems they use than would be expected simply on the basis of those carriers' shares of service in the local markets where agencies are located. This 'halo effect' appears to result from factors other than incentive or override commissions paid to travel agents, since it has been found to persist even in the absence of agreements by CRS-owning airlines to pay these commissions (US Department of Transportation, 1988). Instead, it appears to stem partly from subtle forms of bias in favour of the CRS-owning or 'host' airline that continue to characterize the structure and operation of each system. In addition, each CRS appears to provide more convenient access to – and more accurate – information on fares and seat availability on flights operated by its airline owner, as well as superior reliability in recording reservations for travel on its own flights.

As a result, ownership of a CRS continues to allow an airline to induce travel agents to divert some passengers who would otherwise travel on flights operated by their competitors to their own flights. This diversion generates additional revenues for CRS-owning airlines (and presumably profits, since it is unlikely that costs increase commensurately), at the expense of their competitors. Although there is some tendency for the resulting transfers among airlines to offset one another, on balance there continues to be a substantial transfer of revenues to the carriers that own these systems from other airlines, as well as to owners of the most heavily used systems from owners of smaller CRSs. Further, the existence of a halo effect may enable airlines that own the largest CRSs to expand further the number of travel agents who subscribe to their systems, since the additional airline revenue it produces raises their financial reward from persuading agents to switch to their systems, thereby allowing them to cross-subsidize more favourable subscription terms to new agents.

Fees for services provided by CRSs

Second, CRSs may continue to influence airline competition by allowing their owners to raise rival airlines' costs for obtaining reservation services above their own, and to benefit financially as a result. Two factors allow these systems' owners to set the fees they charge other airlines for recording passenger bookings above the costs they incur in doing so. With an increasing share of airline travel now sold by computerized travel agencies, virtually all major airlines regard it as necessary to accept reservations from agents equipped with even the smallest CRS, which is used by about 5% of US travel agencies. Because even the largest airlines apparently believe that significant revenue losses would result from their refusal to pay the booking fees established by any CRS (thereby making them unable to receive reservations through that system), they exercise little direct bargaining power over the fee levels set by these systems' owners. Airlines' ability to guide bookings on their flights through a system that offered reduced fees – another potential source of bargaining power over fee levels – is also limited by their inability to influence potential travellers on their flights to patronize agents equipped with such a system, or to persuade agents to equip themselves with it.[26]

As a consequence of these limits on other airlines' bargaining power, CRS owners have the ability to charge fees for their systems' services that exceed the costs they incur in processing and storing passenger reservations. One estimate is that the largest CRSs charge booking fees per reservation that are approximately double their airline owners' unit costs for processing them (US Department of Transportation, 1988). This raises the costs to airlines that do not own CRSs of accommodating the substantial fraction of passengers who make reservations for travel on their flights through travel agents (rather than by contacting the airline directly). Even for CRS-owning carriers, their mutual ability to set booking fees above reservation processing and storage costs raises their expenses for serving passengers who make reservations through travel agents equipped with competing systems. As with the halo effect, much of the resulting revenue transfers occur among the CRS-owning airlines, although on balance CRS-owning airlines gain both a cost advantage over their rivals and a substantial financial transfer from them. Further, the airline owners of the largest CRSs enjoy these same advantages over even their competitors who own less widely used systems.[27]

Potential remedies for the effects of CRSs

Computer reservation systems apparently continue to 'tilt' the playing field of airline competition in favour of their owners, and particularly to the further advantage of the largest carriers, who not coincidentally own the systems most widely employed by travel agents. Unfortunately, remedying their influence on competition is likely to require draconian measures. While the influence on airline competition exercised by these systems' persistent halo effects – which stem from their ownership by airlines – is likely to be diluted by the current tendency toward diversification of their ownership among multiple airlines, it will probably be eliminated only by forced divestiture of CRSs from airline ownership of *any* form or degree.

Similarly, the sources of CRS-owning airlines' power to charge fees for reservation services exceeding the costs they incur in providing them are unlikely to be undermined by foreseeable developments. Hence eliminating this avenue for their influence on competition seems likely to require government regulation or even outright prohibition of booking fees, which would probably entail comprehensive restructuring of both subscription fees paid by travel agencies to CRS vendors and airlines' commission payments to agents. Although such measures are thus likely to be extremely controversial and politically unpalatable, as well as perhaps disruptive to the CRS and travel agent industries in the short term, they may nevertheless be warranted by the potential long-run benefits from reducing these systems' apparently pervasive influence on competition among airlines.

Access to airport facilities by potential competitors

In order to establish competing service to a city, an airline requires use of various airport facilities such as passenger lounges and boarding gates, ticket counters, and baggage handling equipment. If access to these facilities is not available to potential competitors at costs comparable to those paid by established airlines, or is unavailable on any terms, entry can be impeded and the benefits of competition reduced. In the US, these facilities are constructed by airport authorities (who are instruments of local, rather than federal, government) and leased to airlines under terms that commonly grant leasing carriers exclusive use of them for extremely long periods of time (often 20 years or longer).[28] Airlines seeking to inaugurate or expand service at an airport where facilities are leased on an exclusive-use basis can gain access to them only by sub-leasing them from carriers who hold the original leases for their use, or by awaiting the construction of new facilities. However, many airport leases provide airlines with veto power over the airport operator's authority to construct additional facilities.[29]

Difficulties experienced by potential competitors
Airlines holding long-term leases on critical airport facilities appear to have offered them for sub-lease to competitors at higher rates than those specified in the original leases. Lease-holding airlines have also occasionally refused to provide use of these facilities to new competitors under any terms, or have offered to do so only on a temporary or interruptible basis, sometimes at substantial 'markups' over their original lease rates (US General Accounting Office, 1989). Dominant airlines at some airports have even attempted to increase the number of critical facilities they control beyond that necessary to operate their scheduled services, in effect 'hoarding' these facilities primarily to prevent their use by competitors (Levine, 1987). Where lease-holding airlines have the power to do so, they have also occasionally used (or threatened to use) it to block construction of expanded facilities intended to provide access by new competitors. Further, local concerns over noise and other environmental consequences of airport use have also severely limited the ability of many large US airports to construct facilities that would permit increased air traffic.

Measures to safeguard access
Ensuring access to necessary airport facilities by competitors seeking to begin or expand service thus appears to be critical to the continued success of deregulation. Doing so will require government agencies responsible for constructing and managing airports to reduce the degree as well as the duration of incumbent airlines' control over access to these facilities. Airport authorities can accomplish this most effectively by reducing the term of facility leases, and by substituting 'preferential use' for exclusive use clauses in leases. Under such a preferential use agreement, the leasing airline would be guaranteed access to a facility – such as a passenger boarding gate – whenever it is scheduled to be needed, but the airport operator would retain the right to assign its use to other carriers during periods when it is unutilized. In addition, airports could reduce incumbent airlines' control over access by new competitors by eliminating the former's control over expansion of airport facilities, as some have already done. Finally, while control over the environmental intrusiveness of airports is a legitimate local interest, it must eventually be balanced against a national interest in maintaining the vigorous competition among airlines necessary to sustain the benefits from deregulation.[30]

Notes

1. With the 1940 reorganization of the CAA to become the CAB, its original jurisdiction over airway and airport development was transferred to the Department of Commerce (and subsequently to the Federal Aviation Administration, now an agency of the US Department of Transportation). The CAA's original authority to regulate airline operations and other factors influencing airline safety was transferred to the Federal Aviation Administration in 1958.
2. No bankruptcies occurred under the CAB's jurisdiction, primarily because the

Board precluded them by arranging for financially trouble carriers to merge with healthy ones.

3. Kaplan (1986) reports that fares remained virtually unchanged in actual dollars between 1950 and 1970, and actually declined by almost 50% in 'real' or inflation-adjusted terms.

4. The latter's authority over airline mergers was subsequently transferred to the Department of Justice at the end of 1988.

5. Whether or not the resulting volumes increase to more or less than exactly the sum of those on non-stop flights will depend on travellers' responses to the additional travel time from A to C imposed by stopping at B and the availability of more frequent services from A to both B and C.

6. Alternatively, a carrier may institute more frequent services on its spoke routes, thus sacrificing some of its potential operating cost savings yet still reducing total travel costs when these are construed to include schedule delays experienced by travellers as a result of imperfect matching between their desired departure times and the carrier's actual schedule.

7. Of course, doing so results in extreme peaking and thus inefficiency in the utilization of passenger and baggage-handling facilities, which imposes some additional costs for the carrier.

8. Such co-ordination is exactly the idea underlying the joint operating and marketing agreements that have been widely established between major airlines and commuter or regional airlines (and, less frequently, between two major airlines).

9. This includes employment in managerial, administrative, professional, technical, and sales occupations; data from US Bureau of the Census (1978, 1979, and 1989).

10. For discussions of the determinants of demand for non-business airline travel, see Morrison & Winston (1985) and Meyer & Oster (1987).

11. The fleet of jet aircraft operated by large US carriers now numbers almost 4,000; in addition, smaller carriers operate another 1,800 propeller-driven aircraft in scheduled service (Regional Airline Association, 1989).

12. Surprisingly little of the popular debate in the US over the effects of the policy appears to recognize this point.

13. Prior to the Domestic Passenger Fare Investigation (DPFI) it conducted during the early 1970s, the CAB focused on carriers' operating expenses per passenger-mile. Since these expenses increase when the percentage of seats that are filled by paying passengers falls, declining load factors provided a basis for airlines to petition for fare increases. Following the DPFI, the Board focused on changes in carriers' operating expenses per available seat-mile (which it attempted to adjust for changes in the average length of flights and for expenses incurred in carrying freight and mail) in determining whether fare increases should be authorized.

14. The Board intentionally set fares above what it estimated to be the costs of service in long-distance markets (those above roughly 400 miles), while restricting fares to levels below average costs on shorter routes in an effort to promote the development of airline service to smaller communities.

15. It is important to recognize that the deregulation of fares began well before passage of the Airline Deregulation Act of 1978. During the spring of 1977, the CAB approved proposals by World Airways, Texas International Airlines, and American Airlines to offer fares that were 30–50% lower than coach fares then in effect. Unlike most previous discount fares that were generally only available to certain groups of travellers (youth, military personnel, etc), these fares were subject only to certain restrictions on their use. By the end of 1977, the Board had approved similar discount fare proposals for virtually all domestic routes served by the carriers it regulated. Thus it is appropriate to consider 1977 (and perhaps even 1976), rather than 1978, as the last year during which the CAB's fare regulation practices prevailed.

16. With the continued transition to an all-jet fleet and the introduction of wide-body jets, the number of seats per domestic flight increased by 54% between 1967 and 1977.

17. Wide-body jet aircraft grew to 19% of the US fleet by the end of 1987, only a slightly higher fraction than a decade earlier. The average number of seats per flight grew by only about 10% between 1977 and 1987. The distance covered by the average flight also continued to increase during the 1977–87 period, but at only about half of the rate at which it grew during the last decade of regulation.

18. During 1977, fares for first-class domestic travel – typically the highest fares paid – averaged 45% above the overall mean fare level (Air Transport Association, 1978), and thus fell in the second-highest category shown in Table 2.3. During 1988, such fares were probably included in the highest category.

19. One possible explanation is that the pricing freedom afforded by deregulation has been used by airlines to shift much of the cost of providing this flexibility to travellers who utilize it. Such costs include holding aside extra seats to ensure the availability of reservations on short notice, many of which remain unutilized, as well as perhaps some of the costs for maintaining sophisticated reservation systems to accommodate the instantaneous reservation confirmation, and schedule flexibility that characterize traditional full-service airline travel. Of course, another possible explanation is that airlines are able to use travel restrictions that have no apparent basis in cost – such as the requirement that travellers stay over a Saturday night – to discriminate between those whose travel demands are insensitive to price and those who are more price-sensitive. If this is the case, airlines may thus be able to charge fares above costs to the former group while maintaining below-cost fares for the latter. Although purely discriminatory price differences cannot be sustained in a competitive environment, the apparently imperfect nature of competition in today's airline market may allow such differentials to be larger than justified by cost differences alone, to arise in more markets, or to persist longer than would be the case in a purely competitive industry.

20. The distance intervals shown in Table 2.3 refer to passengers' origin-to-destination trips rather than to flight distances, so the actual 1988 fares shown in the table account for the effects of increased hubbing activity on flight distances, and thus on carrier costs and fares. Because the increased use of hub-and-spoke routings increases passenger volumes on individual flights (thus reducing unit operating expenses, as discussed previously), but reduces flight distances – thereby raising unit operating expenses – the net effect of increased hubbing activity on variation of fares with distance is difficult to anticipate.

21. The data in Table 2.4 actually applies to points served exclusively by aircraft of larger than 40 seats during 1979. Since unregulated 'commuter' airlines were not permitted to operate aircraft of more than 30 seats until passage of the 1978 Act (which raised this limit to 56 seats), the 40-seat threshold was virtually synonymous with the distinction between regulated and unregulated service during 1979.

22. The US Department of Transportation classifies airports and the communities they serve into four size categories – large, medium, small, and non-hubs – based on the proportion of total nationwide airline passengers boarding flights there. Because these passenger boarding figures include both travellers originating trips and those transferring from one flight to another, the resulting airport and community size classifications are not perfectly associated with population or other determinants of total travel activity, although their relationship is still close. With a few notable exceptions, large hub communities tend to be urban areas with populations over 2 million, while medium hub populations typically range from 500,000 to 2 million, and small hub cities usually have populations from 100,000 to 500,000 (non-hubs are typically less than 100,000 in population). During 1988, 27 communities were classified as large hubs, 35 as medium hubs, and 62 as small hubs; the remaining

communities receiving scheduled service (approximately another 100) were classified as non-hubs.

23. In certain market size categories, the two measures of service changes shown in the table appear to conflict. The increase in direct plus connecting service in each market size category (column 2 of the table) should be at least as large as the increase in direct service alone, yet in the large–large, large–medium, large–small, and medium–medium, and medium–small categories, this is not the case. Since the two measure employ service frequency for June and October 1977, the changes reported in the table should be calculated from roughly comparable baselines. Thus the anomalies between them must be attributable to the fact that the direct service measures is calculated for all routes in each size category and extends through October 1984, while the direct-plus-connecting service measure is based on a sample of routes in each size category and extends through June 1983.

24. The safety-related operating and aircraft maintenance practices of US airlines are regulated by the Federal Aviation Administration, an agency of the US Department of Transportation. Its regulation of carrier practices was not affected by relaxation and subsequent elimination of the CAB's *economic* controls over the industry. The usual logic underlying a hypothesized linkage between economic deregulation and airlines' safety performance is that it may cause them to reduce maintenance and other safety-related spending as part of their efforts to control costs in response to competitive pressures introduced by economic deregulation.

25. The attractiveness of more prominent CRS display may be so large as to induce major carriers to acquire ownership of their joint marketing partners or begin their own regional airline subsidiaries if code-sharing were prohibited, but competing major carriers would at least be provided an equal opportunity to do so.

26. Although airlines could offer to rebate part of the resulting savings to travel agencies that chose to use a system offering reduced booking fees, agencies' opportunities to switch from one system to another are restricted by their long-term contracts with CRS vendors, which also specify extremely high penalties for travel agencies that terminate these agreements.

27. A third advantage that may be realized by CRS-owning airlines is the ability to use their systems to obtain 'real-time' information on passenger booking patterns. Because travel agents use a CRS to make reservations on virtually all carriers' flights, an airline that owns a widely-used system receives a continuous sample of passenger bookings for travel on its competitors' flights as well as its own. In effect, CRS-owning airlines enjoy a technological advantage over their rivals in obtaining current information, which they can exploit to monitor the performance of travel agents in 'steering' travellers toward their own flights, and thus to enhance the effectiveness of marketing initiatives (such as their payment of override commissions to travel agents), again at the expense of their competitors; see Levine (1987).

28. A comprehensive survey of large US airports conducted by the US General Accounting Office (1989) revealed that 83% of all passenger boarding gates at these airports are leased under agreements that will not expire for more than two years, almost two-thirds of which will not expire for more than *10* years. These gates are one of the most vital facilities for an airline wishing to begin or expand service, yet are the facility most commonly leased on a long-term, exclusive-use basis.

29. These agreements are presently in effect at roughly half of the nation's largest airports, including an even higher fraction of those where one or two airlines operate most of the flights (US General Accounting Office, 1986).

30. To whatever extent the fare and service benefits of more intensive competition at an individual airport are received by residents of the city where it is located, the consequences of local restrictions on airport use are also locally borne (although still perhaps somewhat less locally than the advantages from, say, reduced noise). Any

national interest arises from the significance of access to individual airports in maintaining competitiveness throughout the nationwide air transportation system.

References

Air Transport Association, *Air Transport: The Annual Report of the U.S. Scheduled Airline Industry*, Washington, DC, various issues.

Airline Economics, Inc. *The Airline Quarterly*, Washington, DC, various issues.

Bailey, E.E., Graham, D.R., & Kaplan, D.P. (1985) *Deregulating the airlines*, Cambridge, Mass: MIT Press.

Borenstein, S. (1988) 'Hubs and high fares: airport dominance and market power in the US airline industry', Institute of Public Policy Studies Discussion Paper #278, University of Michigan, March 1988.

Call, G.D., & Keeler, T.E. (1985) 'Airline deregulation, fares, and market behavior: some empirical evidence'. In Daughety, A.F. (ed), *Analytical studies in transport economics*, Cambridge, England: Cambridge University Press.

Carlton, D., Landes, W., & Posner, R. (1980) 'Benefits and costs of airline mergers: a case study', *Bell Journal of Economics* 11. 65–83.

Caves, D.W., Christensen, L.R., & Tretheway, M.W. (1983) 'Productivity performance of U.S. trunk and local service airlines in the era of deregulation', *Economic Inquiry*, 21 (July) 312–34.

Eads, G.C. (1972) *The local service airline experiment*, Washington, DC: Brookings Institution.

Fruhan, W.E. (1971) *The fight for competitive advantage: a study of the United States domestic trunk air carriers*, Boston: Division of Research, Harvard Business School.

Gillen. D.W., Oum, T.E., & Tretheway, M.W. (1985) *Airline costs and performance: implications for public and industry policies*. Vancouver: University of British Columbia.

Kaplan, D.P. (1986) 'The changing airline industry'. In Klass, M.W. & Weiss, L. W. (eds), *Regulatory reform: what actually happened*, Boston: Little, Brown & Co.

Levine, M.E. (1987) 'Airline competition in deregulated markets: theory, firm strategy, and public policy', *Yale Journal on Regulation*, 29 393–494.

Meyer, J.R., & Oster, C.V., Jr. (1981) *Airline deregulation: the early experience*. Boston: Auburn House Publishing Co.

Meyer, J.R., & Oster, C.V., Jr. (1984) *Deregulation and the new airline entrepreneurs*, Cambridge, Mass: MIT Press.

Meyer, J.R., & Oster, C.V., Jr. (1987) *Deregulation and the future of intercity passenger travel*, Cambridge, Mass: MIT Press.

Morrison, S.A., & Winston, C. (1985) 'An econometric analysis of the demand for intercity passenger transportation' *Research in Transportation Economics*, 2, Greenwich, Connecticut: JAI Press, 213–37.

Morrison, S.A., & Winston, C. (1986) *The economic effects of airline deregulation*, Washington, DC: Brookings Institution.

Morrison, S.A., & Winston, C. (1989) 'The dynamics of airline pricing and competition' paper presented to the Annual Meeting of the American Economics Association, Atlanta, USA, December 1989.

Oster, C.V., Jr., & Zorn, C.K. (1987) 'Airline deregulation: is it still safe to fly?' In *Transportation deregulation and safety: conference proceedings*, Transportation Center, Northwestern University, pp. 265–93.

Rose, N.L. (1987) 'Financial influences on airline safety'. In *Transportation deregulation and safety: conference proceedings*, Transportation Center, Northwestern University, pp. 295–324.

Stephenson, F.J., & Fox, R.J. (1987) 'Corporate attitudes toward frequent-flier programs' *Transportation Journal*, **11**, 10–22.

Tretheway, M.W. (1989) 'Frequent flyer programs: marketing bonanza or anti-competitive tool?' *Journal of the Transportation Research Forum*, **30**, 195–201.

US Congressional Budget Office (1988) *Policies for the deregulated airline industry*, Washington, DC: US Government Printing Office.

US Department of Transportation (1988) *Study of airline computer reservation systems*, Report DOT-P-37-88-2, Washington, DC: Office of the Secretary of Transportation.

US General Accounting Office (1985) *Deregulation: increased competition is making airlines more efficient and responsive to consumers*, Report GAO/RCED-86-26. Washington, DC: US General Accounting Office.

Viton, P.A. (1986) 'Air deregulation revisited: choice of aircraft, load factors, and marginal cost fares for domestic travel' *Transportation Research*, **20A**, 361–71.

CHAPTER 3

The regulation and deregulation of Australia's domestic airline industry

Peter Forsyth

3.1 Introduction

During 1990, Australia will have partially deregulated its domestic airline industry. An extensive system of regulation will be removed, and it will become permissible for new airlines to enter the industry. International airlines which serve Australia, including the Australian government-owned Qantas, will not be permitted to enter domestic routes however. In this chapter, the system of regulation, the transition to deregulation, and the likely performance under deregulation will be considered. Attention is concentrated upon domestic aviation, though reference is made to international aspects, and especially, those which arise in the South Pacific context.

The chapter begins with a brief description of the industry and the historical background to regulation in section 3.2. In section 3.3, the system of regulation, the Two Airline Policy, is described, and in section 3.4 performance under the policy is analysed. Then, in section 3.5, the pressures leading up to deregulation are examined. In section 3.6, the scope of the changes being made is outlined, and in section 3.7, the likely performance under deregulation is considered. The chapter concludes with an assessment of deregulation in section 3.8.

3.2 Description and historical background

The Australian aviation industry

Australia is a large and isolated country, and like Canada, it is sparsely populated. Distances between major cities are large, and there is scope for considerable reliance on aviation. Australia consists of five major cities over one million in population, a few major tourist areas such as Cairns on the Barrier Reef, and few large cities in between. There is an extensive domestic airline network. Some comparisons between Australia, the US and Canada are made in Table 3.1.

Table 3.1 Population and the domestic airline industry.

	Total population (millions)	Estimated origin – destination trips (millions)	Passenger kilometres (millions)
Australia 1984–85	15.7	10.1	10,200
US 1980s	238.7	242.0	393,200
Canada 1985	25.4	11.9	17,500

Source: May Report, (1986) vol 1, p 68.

Australia's isolation means that the international and domestic airline systems are rather different. Domestic services are, in the main, short hauls operated by small to medium-sized aircraft. International services are very long hauls operated by wide bodied aircraft. Australia's international carrier, Qantas, has twice the average stage length, and nearly twice the average aircraft size of most other airlines. There are few short haul international routes – those that exist are in the South Pacific region, and apart from the routes to New Zealand, most are low density routes. There has been a separation of international from domestic aviation enforced by regulation – however, if this regulation did not exist, the two markets would remain separate except for routes in the South Pacific region.

As Table 3.1 indicates, the Australian market is a small one by US standards, and it is somewhat smaller than the Canadian market. There are two trunk airlines, Australian Airlines and Ansett Airlines. These are about the same size as the US short-haul national airlines (see Table 3.8) and they are considerably smaller than the US majors. The trunks account for about 85% of total traffic. There is a fringe of regional airlines, (see Table 3.2) – all of these are now owned by Ansett, except for Air Queensland which was purchased by Australian Airlines and subsequently closed down. The two trunk airlines have a very similar share of the traffic (see Table 3.3) – this is mainly due to the fact that they

Table 3.2 Airline characteristics 1984/5.

Airline	Revenue passenger km (million)	%	Passenger load factor (%)
Australian Airlines	4,907	43.6	74.2
Ansett	4,835	43.0	75.2
Trunks	9,742	86.6	
East West	622	5.5	66.8
Air Queensland	64	0.6	60.4
Air NSW	260	2.3	59.5
Airlines of South Australia	36	0.3	58.9
Ansett Western Australia	462	4.1	62.2
Ansett Northern Territory	57	0.5	68.3
Regional	1,501	13.3	
Total	11,243	100.0	

Sources: *May Report*, (1986) vol. 2: 73 and International Civil Aviation Organization, and Department of Aviation, *Air Transport Statistics*.

Table 3.3   Traffic proportions and load factors, 1961–85/6.

| | Australian Airlines | | Ansett Airlines | |
	Passenger % of traffic	Load factor	Passenger % of traffic	Load factor
1961	46.0	—	39.9	—
1965	45.5	—	42.1	—
1970	45.7	67.5	43.1	64.1
1975	45.2	64.3	44.5	67.0
1980/81	—	76.3	—	74.2
1985/86	41.4	78.0	43.5	74.6

Sources: *Domestic Air Transport Policy Review*, report vol. 1.
Tables 14.1 and 14.2, Departments of Transport and Aviation, *Air Transport Statistics*.

have been regulated so as to have the same amount of capacity. Load factors have been high by international standards (again reflecting capacity regulation) and recently they have risen. The Australian market consists of a small group of quite high density routes (see Table 3.12), plus a fringe of low density routes served by the trunk and regional airlines.

Table 3.4   Passenger traffic ('000 passenger embarkations).

	Trunk	Total
1974–75	7885	9787
1978–79	8821	11,323
1983–84	9330	11,551
1987–88*	11,542	13,648

* Break of series
Sources: Bureau of Transport Economics (1985) p. 11 and Bureau of Transport and Communications Economics, *Indicators*, quarterly, and Department of Transport, *Air Transport Statistics*.

The Australian domestic market has been growing modestly over the past 10–15 years (Table 3.4). It has been growing rather less rapidly than international traffic from Australia, and much less rapidly than international traffic to Australia. It has also been growing significantly less rapidly than US traffic in the post-deregulation decade. This is not surprising when one considers how real fares have moved. As Table 3.5 shows, real fares have remained relatively unchanged for a very long time. Yields per passenger kilometre would show a slight declining trend, since there has been some increase in the use of discount fares. This increase is modest, and discount fares are neither as low nor as widespread as they are in the US. The pattern of falling real fares, and increased availability of discount fares, which has been the experience in the US since deregulation, and which has also been the experience of Australian international fares, has not been repeated within Australia.

Table 3.5   Real air fares 1960–87, International fare index.

| | International fare index | | Domestic |
	Economy fare	minimum fare	real fare index
1960	–	–	102.6
1965	–	–	99.6
1970	–	–	95.1
1971	100.0	100.0	–
1975	83.7	69.9	88.9
1980	85.1	46.1	97.1
1985	106.6	53.6	131.9
1987	103.5	48.9	128.0

Sources: *Domestic Air Transport Policy Review*, vol. 11
appendixes (1979): 118, May Report (1986) vol. 1: 388,
Australian Bureau of Statistics, CPI and Bureau of
Transport Economics, *Transport Indicators*, numerical
series.

A description of the domestic airline industry in Australia, written in 1969, would not be very inaccurate for 1989. The size of the industry has increased, though not dramatically. The real level of core fares has remained much the same, though there are more discounts available. The market structure has remained the same, with two nearly identical airlines accounting for over 80% of the traffic. The core networks they fly are much the same as before, though they have shed low density routes. This lack of change can be attributed to the detailed regulatory system which the industry operates within.

The historical background

For several decades, Australia has had a unique system of domestic airline regulation – the Two Airline Policy. This policy has resulted in two very similar airlines, which have operated under very strict regulation (and a measure of self-regulation). The policy has no strict parallel – perhaps the closest system was that in Canada, where a public and a private airline operated with a fringe of regional airlines; while outwardly similar, the Canadian case was quite different in the closeness of the regulation. At its apotheosis, in the 1970s, the Two Airline Policy resulted in virtually all economic aspects of the airlines being regulated (normally to be the same) by the government (which reflected the airlines' wishes in the main) or jointly regulated through a 'rationalization' system of self-regulation.

Partly because of its uniqueness, the Australian airlines have been subjected to a good deal of scrutiny in the literature. The pre-Two Airline Policy period was studied by Hocking and Haddon-Cave (1951). Since then different aspects have attracted the attention of experts – Corbett (1965) and Wettenhall (1962) examined the political and administrative aspects of the policy, Poulton and Richardson (1968) the legal aspects, Blain (1983) some of the industrial relations aspects and Brogden (1968) gives a history of the formative years. Even though it is a form of economic regulation, there has been no full-length study of the economic aspects, Hocking (1972), Forsyth and Hocking (1980) and Kirby (1981) are shorter monographs on the subject. A recent discussion of some of the economic aspects is contained in Centre for Independent Studies (1984).

Apart from these independent studies, there are several government reports which discuss the workings of the policy. In 1979 there was the Domestic Air Transport Policy Review (Dept of Transport, 1979). In 1981, the report of the Holcroft Committee into domestic air fares was published (Holcroft, 1981). Most recently there has been the May Committee Inquiry which reported in 1986 (May, 1986). There have been reports analysing state or territory airline regulation in NSW (Riley, 1986), WA (1982) and Gallagher (1980). The Bureau of Transport Economics has published reports on the airline system (BTE, 1985). If the number of times that the policy has been referred to experts for diagnosis is any indication of the doubts or dissatisfaction with it, then it cannot be said to have been very satisfactory in the community's eyes. Because the policy has been so well covered, it is not necessary to go into the details of its history, legal background or politics here.

Before the Two Airline Policy

At the end of World War II, Australia had a large privately-owned airline, with a fringe of smaller operators. The major airline, Australian National Airways, (ANA) had done well out of its wartime operations. At the time, a Labor government was in power, and it saw nationalisation of key industries as an important means of control of the economy. It decided to attempt to create a nationalised airline industry. The government, however, was thwarted from forcibly nationalising the airline by a successful constitutional challenge. It responded to this by setting up its own airline in 1946, Trans Australia Airlines (TAA) now Australian Airlines, in an attempt to compete the private airline out of business.

This airline was not required to earn a profit or cover costs, and partly by cut-price competition, it soon managed to gain inroads into ANA's markets, and by 1948, the owners of ANA were suggesting to the government that the two airlines merge to form a new airline, with public and private shareholdings. The government rejected these proposals and continued to allow TAA to grow at ANA's expense.

Birth of the Two Airline Policy, 1950-7

In 1949, the Labor government was replaced by a conservative administration. This government had opposed Labor's nationalisation proposals, and originally proposed privatisation of TAA. It changed its mind, but soon set about creating a more equal environment for the airlines to operate in. This involved reductions in charges to ANA, and increases in charges to TAA, which was from then on expected to operate at a profit. In 1951 it stated that there was to be a 'Two Airline Policy', and that it would ensure that the public and private airlines both survived, and competed against each other. It was of the view that scale economies existed in the industry, but that these would be reaped by two large airlines, and that competition between the two would result in better performance than that which would be achieved by a monopoly.

In the 1950s, there was a 'Two Airline Policy' in name only. There was little in the way of restrictions on other airlines which wished to compete, and the government allowed these to import aircraft. One airline, Ansett, became an aggressive competitor and offered cut-price flights on major routes. During this period the main private airline encountered a series of crises; for example, as a result of inappropriate equipment choices (it chose long range DC6 aircraft whereas TAA chose faster, shorter range, and smaller Viscounts). Its financial performance deteriorated, and it was constantly requesting (and demanding) help from the government. In 1957, Ansett offered to take over the ailing ANA, on condition that the government introduce tighter regulation of the industry (designed to prevent other airlines from doing what Ansett had been doing).

Consolidation of the policy, 1957-7

The government agreed to a more tight regulatory environment, and set about creating it. It did this in several ways. It entered into a legal agreement with Ansett to operate and enforce a Two Airline Policy – thus the government itself was constrained from changing the policy later on. It used its powers over imports to stop any other airlines from getting aircraft they could use to compete with the chosen two. It further extended its policy of putting the two airlines on an equal competitive footing. A framework for co-ordinated decision-making (collusion) over routes and fares, the Rationalisation Committee, which had existed before, was strengthened and made more effective. Further changes were made in 1958 when a system of capacity control was introduced. The two airlines were to have equal capacity, and capacity was to be expanded only when there was a demonstrable need for it. Existing capacity imbalances were later removed by exchanges of aircraft – TAA, with its more appropriate fleet, considered that it had been adversely affected by this.

From about 1960, the Two Airline Policy was in place. Little has changed since then, until recently. There have been two almost identical airlines with all the busy routes. Their shares of the traffic have remained very similar. There was a fringe of small airlines, mainly servicing regional markets. Gradually all of these were acquired by the major airlines, mainly Ansett. There are a number of commuter airlines, which have developed considerably in the past two decades. Currently, most of these have commercial and interlining agreements with the two majors, and use their booking systems. One regional airline, East West, remained independent until 1987 – for most of the period it was content with its regional status, but early in the 1980s it began to provide some competition for the majors on selected busy routes. It was tightly constrained by regulation, and its impact was limited.

The policy under scrutiny, 1977-87

The airline system attracted rather little attention until about 1977; since then it has been subjected to constant scrutiny. At different stages, government inquiries have examined the industry (the Domestic Air Transport Policy Review, 1979, the Holcroft Inquiry into Air Fares, 1981 and the May Inquiry 1986). The reports of the latter two inquiries were quite critical of the performance of the industry. In 1981 the government revised the Two Airline agreement, in some ways making the regulation tighter (through tighter restrictions on entry, and price controls) and in others less tight (giving the airlines more scope to compete on price). The agreement was extended for five years, with a period of three years required for notification if the policy was to be ended. After considering the May Report of 1986, the government announced, in October 1987, that the Policy would end in October 1990.

International aspects of the policy

Since the 1940s, Australia's airline policies have enforced a rigid separation between domestic and international aviation. Qantas, a private airline which was nationalised, became the sole international carrier (except for some minor short-haul routes). It was not permitted to operate domestic services. It does fly on several domestic segments (e.g. Melbourne–Sydney, Sydney–Perth) which form part of its international route network. It has always been permitted to carry its own passengers on these route segments (e.g. to carry a Los Angeles–Melbourne passenger from Sydney to Melbourne). For a time (1979–88) it was not permitted to carry other international airlines' passengers on domestic segments.

International airlines from other countries are not permitted to sell seats on domestic segments, though they are able to operate on these as part of their international flights. In this respect, Australia's policy is similar to those of most other countries. The policy does have particular significance for airlines of nearby countries, especially New Zealand. Such airlines, along with Qantas, are not permitted to construct networks which have international and domestic segments, and to sell their empty seats on those segments. Thus, neither Qantas nor Air New Zealand could operate Auckland–Melbourne–Sydney–Auckland and sell seats between Melbourne–Sydney, even if it made commercial and operational sense to do so. Ansett operates domestic services in Australia and through a subsidiary, Ansett New Zealand, to New Zealand but it cannot link up its services.

Qantas, and the Ansett New Zealand-based airlines are ones most affected by the separation of domestic and international aviation. Most domestic flights are operated by small to medium-sized short-haul aircraft, and most international flights are operated, by virtue of Australia's geographic isolation, by wide-bodied long haul aircraft. Most international airlines flying to Australia would have little interest in or suitability for operating, except in a marginal way, on domestic routes, unless they were to establish an Australian-based subsidiary. Qantas has occasionally expressed an interest in operating domestic flights, but currently is not very interested (it probably sees the segmentation of markets as operating to its advantage). The airlines most constrained are Ansett, which would like the ability to fly on international routes, and Air New Zealand, which would like to have the same opportunities in Australia as its competitor, Ansett, enjoys in New Zealand.

3.3 The Two Airline Policy

Australia's Two Airline Policy is of interest for several reasons. It is an exceptionally tight system of regulation that has withstood pressures for competition for nearly three decades. It is atypical in that the regulation takes the form of a contract between the regulated and the regulator; this has the effect of binding the government, for a period, to maintain the regulation. It has resulted in the division of the airline market between two airlines which are constrained to be very similar.

Entry restrictions

The keystone of the Two Airline Policy has been the control of entry, and this has been exercised through import controls. The federal government has undoubted constitutional power to restrict imports, and to support the Policy, it has restricted imports of aircraft. In order to compete, it is necessary for an airline to have aircraft, and this means importing them (the existing airlines are unlikely to sell their aircraft to potential competitors). The Two Airline Policy has been directed mainly towards passenger traffic, and in 1981, air freight was deregulated. Prior to this, some air freight services other than those operated by the two major airlines existed, but even these had difficulties in securing modern aircraft (for fear that these could be converted into passenger aircraft). In the early 1980s, the then independent regional airline, East West, attempted to import aircraft suited for trunk routes, but by the time it was taken over, it had made little progress. Even if they had not been restricted by other regulations, it would not have been possible for new airlines to enter trunk markets.

Capacity and route restrictions

Capacity also has been strictly regulated. The two trunk airlines have not been able to import at will – they have only been permitted to import aircraft when both of them have been expanding capacity by the same amount. This has meant that, for much of the currency of the Policy, the airlines operated virtually identical fleets. While not strictly constrained to do so, they were under pressure, until the early 1980s, to purchase the same aircraft. When new aircraft were purchased, the Ansett and TAA (Australian) (aircraft would arrive in Australia on the same day. (This broke down when Australian Airlines purchased Airbuses and Ansett purchased Boeing 767s.) Capacity control has been exercised through entry permits; the government as regulator would make an estimate of future traffic, and with an assumed load factor calculate capacity needs – it would then permit airlines to import aircraft to meet the capacity requirements. Only rarely (for example in the 1982/3 recession) did the airlines find themselves with excess capacity.

A system of route controls also existed. There has been a defined set of routes, mainly between capitals and major tourist destinations, defined as 'trunk routes', and these have been (effectively) reserved for the two major airlines. There exist non-trunk routes; many of these are intra-state routes, which can also be subject to stage regulation. These are usually monopoly routes. The major airlines serve some of these, but mostly they are reserved for the regional airlines (most of which are owned by Ansett). There has been little scope for regional airlines, such as the then independent East West, to compete with the major airlines. They could provide limited competition – East West flew indirectly from Sydney to Melbourne but had to use slower turboprop aircraft. It did not make a major impact.

Within these route controls, the major airlines had considerable freedom. They could develop new routes, and change their network. They were able to develop hubs at the major cities, such as Melbourne and Sydney, and choose whether to serve cities by direct or indirect routes.

Price controls

The airlines have all been subjected to price control. Originally, this was done by direct regulation by the department. Over time, a fare formula was developed, which set maximum economy (coach) fares according to a fixed amount and an amount related to distance travelled. Other fares, such as first class and discount fares, were set with reference to economy fares. Until the mid 1970s there were few discounts, except for those given to particular categories of travellers, such as children and pensioners. After an inquiry into air fares (Holcroft, 1981) a specific fare regulator, the Independent Air Fares Committee, was established. This modified the formula, and set more explicit procedures for determining allowable fares. Essentially it has practised rate of return regulation, though it has attempted to monitor efficiency and ensure that cost padding does not take place. Non-trunk airlines are also subject to fare regulation. One of the potentially important changes made at the time of the introduction of the committee was to allow greater flexibility to the trunk airlines to offer discount fares; they have used this option rather sparingly.

The scope for collusion

Within a framework such as this, it should not be surprising if collusion between the major airlines developed. In fact, it was officially sanctioned through a device called the 'Rationalisation Committee'. This was a body formed by representatives of each of the airlines and the government. It was set up in the early days of the Policy to provide a forum for the 'rationalisation' of fares, routes and schedules. In the heyday of the Policy, in the 1960s, the airlines operated very closely together. In the past decade or so, the airlines have become more independent, and have competed against each other in various ways (frequency, in-flight service, advertising and, in some cases, availability of discount fares). Most 'rationalization' provisions were abolished in 1981, though some provision for consultation remains.

Parallel scheduling

Granted the extent and detail of regulation under the Two Airline Policy, it is interesting to note that perhaps the most obvious and curious feature of the Australian airline system, parallel scheduling, is not, at least directly, a result of regulation. Australian schedules are inconvenient for passengers since, on a large number of routes, flights by the two airlines take place at the same time (for an example, see Table 3.6).

Table 3.6   Selected airline schedules, October 1987.

| Route | | Airline | |
		Ansett	Australian
Adelaide–Perth	Daily	10.30am	10.35am
	M,W,F	5.35pm	6.30pm
Adelaide–Alice Springs	Daily	8.40am	8.45am
	Daily	1.35pm	
Adelaide–Canberra	Monday		1.15pm
	Sunday	6.40pm	6.35pm

Source: Airline timetables

Where flights are frequent, as they are between Sydney and Melbourne, this is of no great consequence, but where flights are infrequent, as they might be between smaller capitals and tourist destinations, the fact that the two flights per day leave within ten minutes of one another imposes a clear cost on the passengers. A measure of this cost can be obtained by noting that if one, rather than two, airlines operated the route, it could do so with one, rather than two aircraft, and enjoy the cost savings obtained by larger aircraft.

In fact this scheduling pattern is the outcome of competition, not regulation (at least directly). Regulation has created two very similar airlines, and their responses to scheduling problems will be similar. The process in operation is akin to that noted by Hotelling (1929) – (see Hocking 1972 and Gannon, 1979). If there is a preferred time for travel, in the sense that there are more passengers who wish to travel at a particular time than at others, then both airlines will schedule their flights at this time. In this way, they both obtain half the traffic. If

Table 3.7   Percentage of parallel flights.

| | Frequency of weekly undirectional routes | | | | |
	0–25	26–50	51–100	Above 100	All routes sampled
December 1977					
Percentage of parallel flights	73.0	84.0	75.8	86.4	80.6
Number of routes	11	5	10	4	30
March 1985					
Percentage of parallel flights	60.4	68.8	78.2	87.5	75.8
Number of routes	3	14	7	6	30

Source: *May Report*, (1986) vol. 2: 48.

one scheduled at this time and the other at the next preferred time, travellers would be better served, but one airline would carry more passengers. It is in the interests of this other airline to reschedule its flight at the more popular time, and increase its share of the traffic. Over time, there have been attempts to lessen the extent of parallel schedules through pressure from the regulator and agreement between airlines, though these have not been very successful (see Table 3.7). A factor which does appear to have made a difference has been the purchase of different types and capacities of aircraft in the 1980s.

Australia and America: differences and similarities

The Two Airline Policy is a quite different form of regulation from the pre-regulation US system. As a result, the impact of deregulation is likely to be different. Briefly, the US system regulated entry to routes, and, to an extent, into the industry (though there were many firms there already). It regulated fares, and through these, there was an attempt at regulating profits. Except rarely, it did not regulate the amount of capacity on a route or overall. Routes differed – on some, there was only one airline, on others many. It was difficult, until immediately before deregulation, to offer discount fares, and there was little incentive to do so. Schedule competition, whereby airlines scheduled additional capacity as a competitive weapon and drove down load factors, was a characteristic of this system.

In Australia, capacity is regulated, and load factors are high (though they are not exactly comparable with US load factors because Australian airlines more often cancel poorly-booked flights and this raises load factors). Fares are regulated, though there is more scope to offer discount fares (though equally little incentive). Profits are virtually guaranteed, though they are loosely controlled. Entry by new airlines is virtually impossible. In the US, airlines had an incentive to be cost-efficient since fares were regulated on an industry-wide level – it was possible for an efficient airline to keep its profits.

In Australia, the industry consists of two very similar airlines – while in theory it would be possible for one to be much more profitable than the other, this would invite changes in policy to the detriment of the more efficient airline.

The evidence is that the Australian airlines are less cost-efficient than comparable US airlines (see section 3.4).

The upshot of this is that the nature of the gains from deregulation could be quite different. In the US, networks and service changes were important – in Australia they are likely to be less so. In the US, widening of the range of discount fares was significant – in Australia, the range may not widen very much, but the emphasis on individual fares could change, i.e. low fares may become more freely available. The lower efficiency of Australian airlines would mean that the fall in real average fares could be greater. There are not likely to be the same gains from increased load factors.

3.4 Performance under the Two Airline Policy

For many years there was little serious criticism of the Two Airline Policy and such criticism as there was centred around the obvious problem of parallel scheduling. The outward performance of the airlines was good. They operated networks which were quite convenient for passengers, though there was some pressure for direct flights between centres served indirectly. Their general service standards were high, though passengers made the usual complaints about inflight catering. Airports were not congested, and the on-time performance of the airlines was good. Above all, they had an exceptional safety record. There have been no fatal accidents involving jet aircraft since they were introduced, and there have been no fatal accidents involving trunk or regional airlines since the 1960s.

The high quality of service was purchased at a price. This became apparent over the 1970s and 1980s when the cost of international travel, to or from Australia, fell appreciably in real terms, but the real cost of domestic travel remained much the same (see Table 3.5). While comparisons of fares per kilometre between domestic city prices in Australia and between Australia and the US or Europe were misleading, comparisons between long-haul domestic routes and Australia–New Zealand routes were not. Comparable distances on New Zealand routes could be travelled more cheaply, a point that was often made by residents of Western Australia.

Over the past decade there has been detailed scrutiny of the performance of the domestic airlines, and sufficient evidence has been gathered for some generalizations about it. It is convenient to separate out different aspects of efficiency, and consider allocative efficiency, cost or productive efficiency, and other aspects of efficiency.

Allocative efficiency

By allocative efficiency we mean here the choice of output levels, quality levels and prices of the different services which the airlines provide. The question is whether the airlines are providing what the market wants.

Table 3.8 Performance indicators, 1983–84.

Airline	Passenger numbers ('000)	Revenue passenger kilometres (million)	Operating expenses per available tonne km	Operating expenses per revenue pass. km	Passenger load factor (%)	Average aircraft utilization (hrs)	Revenue passenger km per employee ('000km)
Australian	4,302	4,267	89.8	13.6	73.6	2,633	545
Ansett	4,367	4,279	74.6	11.8	72.9	2,556	582
Air California	3,565	2,304	53.6	11.8	56.8	2,640	1,251
Frontier	6,002	6,633	50.1	10.2	60.2	2,544	1,283
Ozark	4,842	4,238	49.7	11.6	54.7	2,631	1,066
Pacific Southwest	8,097	5,012	45.4	10.4	54.9	2,762	1,372
Southwest	10,790	7,030	30.1	6.8	60.8	3,787	2,074
Average of short-haul nationals	5,987	5,041	41.4	9.7	58.3	2,748	1,465

Source: BTE (1965): Tables 3.16 and 3.17.

By some indicators the Australian airlines appear to do quite well. This is apparent when load factors are considered. During the operation of the Two Airline Policy, load factors have been high by international, and by North American, standards. In the years to 1980, passenger load factors were rarely below 65%, since then, they have averaged above 70% (see Table 3.3). The recent increase in load factors may be due to better use of the price structures e.g. using discount fares to fill up aircraft. The high load factor has been partly caused by regulation – the capacity regulation operated to ensure that a target load factor of around 65% was reached.

It is also worth noting that load factors are not strictly comparable between different airline systems, such as those of Australia and the US, because the scheduling systems are different. In Australia, a flexible scheduling system is used – a timetable is operated, but there are many flights operated which are not listed in the timetable. Thus the timetable indicates a basic level of service – as demand becomes clearer, extra, non-scheduled services are operated. Sometimes scheduled flights are cancelled because of poor demand. Thus, supply is more closely matched to demand than with a rigid timetable system. This enables a higher load factor to be achieved with a given degree of schedule convenience to the passenger (see Gannon, 1979). One might expect this system to result in lower aircraft utilization, though in fact, in Australia, utilization is high (see Table 3.8).

The air fare structure in Australia is basically an efficient one which gives airlines sufficient flexibility to fill flights. The core fares are the economy (coach) fares on which the majority of passengers travel. In addition there are first class and business class fares, and two types of discount fare. The first type of discount is that given to a particular group, e.g. students, and the second is that given to passengers who meet certain restrictions. Thus there are discounts for advance booking and for being prepared to travel on flights nominated by the airline (which gives the airlines greater scope to match demand to capacity). There are also standby fares. The size of the discounts varies from small to moderate (45%). It is difficult to tell how effectively the fare structure is used to fill capacity, granted that capacity has been controlled, and that a flexible scheduling system is used.

The major problem with the air fare structure is that there is a restricted offering of discount fares, and, arguably, the average quality of service is too high. While discount fares are listed, they are not easy to get (nor are the discounts large). In 1983/4, the revenue dilution (shortfall of actual yield from yield with economy fares) on Ansett was 14.9% and on Australian, 25.9% (BTE, 1985: 45). In 1984–5 the proportion of trunk passengers on discount fares was 42%, whereas the proportion in the US, for all domestic routes, in 1985, was over 80% (May, 1986, vol. 2: 42). As a result, there is a smaller proportion of the Australian population which travels by air – in 1985 28% of the US population travelled by air, whereas 21% of the Australian population made inter-state trips (most trips are inter-state) (May, 1986, vol. 2: 36).

The picture then is of a system which serves business markets well, but which is less good at serving leisure markets. In the regulatory environment, the airlines have not needed to seek out the discount traveller, and they have not done so. Even with no change in productive efficiency, lower fares could be offered if airlines lowered service quality (e.g. operated with more seats per aircraft) and there may be some scope for higher load factors. While it is difficult to be certain, there is evidence that the regulatory system has resulted in a substantial market of low fare travellers not being served.

Another aspect of the price structure concerns cross-subsidization. Given the way air fares are set by formula, there is the scope for some cross-subsidization. The aspect which has attracted most attention has been that of subsidies from long-haul to short-haul routes. There can also be subsidies from high-density to low-density routes. The Holcroft Inquiry (Holcroft, 1981) examined the question of cross-subsidies in some detail, and the Independent Air Fares Committee has since sought to reduce the cross-subsidy by distance, though it has not addressed the density question. Over the 1970s and 1980s, the trunk airlines have considerably reduced their services on low-density rural routes – these have been taken over by commuter operators. In the discussion on deregulation there has been little assertion of the need to retain regulation in order to support cross-subsidies, and it is not likely that the allocative efficiency losses have been major.

Productive efficiency

By 'productive efficiency' we mean here how low the cost is at which an airline can produce an output of a given quality level. It combines two aspects – firstly whether the choice of inputs is efficient, and secondly, the technical efficiency aspect of how much output can be derived from specified inputs. It essentially involves how low the cost curve is. If costs are higher than they need be in the Australian airlines, there is the potential for gains if deregulation succeeds in forcing costs to the minimum.

Over the past decade, there has been considerable discussion of the productive efficiency of the Austirlian airlines. It is no longer accepted that costs are at a minimum, but there is disagreement about how much lower costs could be. A natural standard for comparison is the US system, which is generally regarded as the most efficient, and which includes many sizes and types of airlines. The comparisons that have been made include those on the basis of econometric studies of costs, and those which have used indicators of productivity, yields and fares. In this section, an attempt is made to synthesize and interpret results rather than present new analysis.

Two econometric studies are those of MacKay (1979) and Kirby (1984). Relying on a study of international and Australian airlines, Mackay suggested that costs of the Australian domestic airlines could be reduced by up to 35%. Kirby undertook a study of US trunk airlines and the two Australian trunk airlines over the period 1971–8, and used a translog cost function. His results were comparable – he also suggested that cost reductions of 35% were possible. Neither of these studies made allowance for service quality and the fact that discount fares with restrictions were more prevalent on US airlines than on Australian. These studies do suggest that substantial gains in efficiency could be achieved.

Another approach is to use indicators, such as labour productivity and fare yields to make an assessment of possible cost reductions. To make useful comparisons, it is necessary to compare like with like, as far as is possible. The most meaningful comparisons would be between Australian airlines and US airlines of similar size, route density and stage length. The Australian airlines are similar to the US short haul national airlines, not the major airlines. In Table 3.8 a number of indicators are given for the Australian airlines and five similar US airlines, along with the average for short haul nationals.

If labour productivity (revenue passenger kilometres per employee) comparisons are made, the Australian performance looks very poor. The Australian airlines productivity is not much above one third of the average, and only slightly above half that of the lowest productivity of the airlines listed. These figures are misleading, since US airlines of this size are able to, and do, contract out more services than the Australian airlines. Australia's isolation, along with an emphasis on self-reliance and an attention to safety standards, have meant that most maintenance is done in-house. However, this can explain only some of the difference, and it is the case that manning levels for many tasks have been high. This is evidence by the fact that in the wake of the 1989 pilots dispute, both Australian airlines plan to operate their services in future using considerably fewer pilots.

Comparison of yields can also be used as an indicator of efficiency if profit margins are comparable (which they are). However it is preferable to compare operating expenses per unit output, and this is done in Table 3.8. Expenses per available tonne kilometre are higher in Australia than in the US, though expenses per revenue passenger kilometre, are only slightly higher (except for the low cost airline, Southwest). Granted that carrying extra passengers by increasing load factors will increase costs, though not proportionately, the best guide to efficiency would lie between these two indicators. On this basis, it would seem that there is scope for some cost reduction in Australia.

Two qualifications need to be made. Firstly, the average service quality of the Australian airlines may be higher – there are fewer restricted discount fares. Secondly, input prices differ. In Table 3.9, input price indexes are given for the

Table 3.9   Airline input prices US, UK, Australia (in US$ terms).

Country	Input price index 1980	Input price index 1984	Input price index 1985
US	100.0	116.7	119.7
UK	89.0	77.5	76.8
Australia	89.0	98.8	87.6

Source: Forsyth (1985), Table 1.

Table 3.10   Average airline fares, 1,000 km US, UK, Australia (in US$).

	Economy fare, 1985 domestic input prices	Economy fare, 1985 US input prices	Excursion fare 1984 US input prices
US	158.43	158.43	93.96
UK	174.95	262.25	126.05
Australia	108.85	137.48	95.33

Source: Forsyth (1985), Tables 4, 5.

US, Australia and the UK. The Australian airlines pay less for their inputs, and thus a comparison of costs per unit output underestimates the potential cost reduction.

Finally, fares can be compared. This is done in Table 3.10 for a sample of economy and excursion fares in Australia, the US and the UK. Australian economy fares in 1985, for 1,000 km trips, were lower than US fares. Once an adjustment is made for input prices, the gap is narrowed, but not eliminated. This evidence appears to contradict that of the previous comparisons, though there are reasons for this. By 1985, coach fares had ceased to be important in the US, and most travel was at fares below them, whereas most travellers in Australia used economy fares. When a comparison is made of excursion fares, the levels are about the same (when evaluated using US input prices). These fares are readily available in the US, but difficult to obtain, and only used by a few travellers, in Australia. They are (more or less) the minimum fares available in Australia, whereas in the US, they are about mid-price fares. In the US, a significant proportion of passengers (perhaps 30–40%) would travel at fares below these excursion fare levels. Granted this, the average fare paid for this distance in the US, after allowing for different input prices, would be about 15–20% lower than that paid in Australia, though on average the US traveller would face more restrictions and perhaps, would travel in aircraft with more seats, though lower load factors. This would be broadly consistent with the impression which emerges from consideration of expenses per unit output or revenue yields.

In summary, there seems evidence that the productive efficiency of Australian airlines is less than that of comparable US airlines. Labour utilization could be substantially improved, and this might lead to cost reductions of the order 10–20%. As deregulation approaches, the airlines are taking steps to reduce their use of labour. The possible gains in efficiency do not appear as great as those suggested by the econometric studies.

Other aspects of efficiency

There are several, less important, aspects of efficiency which are worth noting. Parallel scheduling has been discussed in some detail. It is a clear aspect of inefficiency, since either, for the same cost, more travellers could be given better flight schedules, or the same schedules could be operated at lower cost. The seriousness of this has declined somewhat, e.g. on low density routes the incidence of parallel flights declined between 1977 and 1985 (May, 1986, vol. 2: 48).

Other aspects of inefficiency arise from the distinctions created between domestic and international routes. Qantas operates several domestic segments but is unable to pick up domestic passengers – such flights often operate with very low load factors. This undoubtedly constrains Quantas's route planning. Other airlines are more constrained, since they can fly only to specified gateways. While this is not likely to be of much importance for, say, KLM, it is a considerable restriction on an airline based in the region such as Air New Zealand. The tight distinction between domestic and international flights means that an efficient route network in the South Pacific region cannot be developed by any airline.

3.5 Pressures for regulatory reform

Gainers and losers

The most clearly articulated support for the Two Airline Policy came from the airlines themselves, especially the private airline, Ansett. The airlines saw themselves as gaining through being guaranteed a comfortable rate of return and a stable environment in which to operate (a quiet life, in other words). Profits were never very high, but losses were rare also. The workforces, and their unions, also gained from the Policy. They were able to secure good wages and working conditions – airlines were unwilling to fight hard against union demands as they knew they could pass on to customers any cost increases incurred. While wages have not been noticeably higher than elsewhere (except for pilots) working conditions have been good – this is reflected in the high manning levels.

Travellers as a group did not gain from the Policy, since they paid higher prices and faced restricted choices of products. They have been dispersed, and have not been very effective as a lobby. There are some groups who take up the case of passengers. There include state governments, which see their states as losing business and tourism, and their voters as paying higher prices for travel. Another group which has been concerned about fares and travel conditions has been the rapidly growing tourism industry. As a supplier of complementary services, this industry is concerned that the Policy has hindered its development. This is a very diverse industry, with few very large operators, and not surprisingly, its influence has been limited. Overall, it is not difficult to explain the persistence of the Two Airline Policy.

Changing attitudes

A possible explanation of the regulatory changes lies with changing attitudes of the various players in the game. The airlines, and in particular the private airline, have altered their assessments of regulation and deregulation. The experience of US deregulation had a considerable effect on their assessments. It became obvious that it was possible for incumbents, with good networks and established names, to survive and prosper in a deregulated environment. They did not need to fear new entrants with lower costs, since any advantages these might have would be short-lived. It might be possible to earn higher, though more variable, profits under deregulation, since the airline would not be held back by regulation from entering new markets or taking some of its rivals' market. It was also clear that established airlines could hold on to resources, such as preferential access to airports and airport terminals, which would give them a clear competitive advantage over new entrants. Deregulation, especially one which was loaded in favour of the incumbents, was becoming an attractive option.

Another element is quite important in explaining Ansett's attitude. In earlier years, it had regarded the Policy as being a guarantee against predatory behaviour by a government-owned firm with access to subsidies. As the experience of the 1940s illustrates, this possibility was quite real. However, over time the emphasis of governments of all political persuasions had moved more towards making Australian Airlines a fully commercial operation, and against using it as a backdoor means of nationalisation. In fact, it may well be that Ansett would now prefer to compete with a government than a private firm, if that government firm is expected to be fully commercial, yet perhaps hampered by the restrictions that would be entailed by bureaucratic oversight. Since 1980, Ansett has switched from opposition to, to support for deregulation – the government airline has not taken a strong line on the regulatory environment.

Other groups have not changed their position. Unions are still, in general, opposed to deregulation, though they did not fight very effectively against it. This has possibly been because the union movement as a whole has been prepared to impose restrictions upon its members in return for concessions from the government. If anything, the success of the tourism industry, and its role as an earner of foreign exchange, has strengthened its voice.

Technological change

Regulatory change sometimes comes when the cost conditions of the industry alter. In the airline industry, perhaps the most significant technological change in recent years has been the development of computer reservations systems. These give an advantage to those airlines which possess them – and both the major Australian domestic airlines operate such systems. Thus, new entrants would find it difficult to break into markets unless they either developed a system of their own (a risky, expensive exercise) or obtained access to the existing systems (under terms set by their competitors). Thus the incumbent airlines may not fear for their prospects under deregulation.

Was regulation breaking down?

Often deregulation comes when the existing regulatory structure is breaking down. For example, in the financial sector, many regulations designed to prevent competition were not working, and banks found themselves constrained in some respects though new competitors were not (see Harper, 1985). In the case of domestic aviation, this was not so – the regulation was proving quite robust. In the early to mid-1980s, the regional airline East West was making some inroads into the markets of the two major airlines. These inroads were not affecting the airlines severely, however, and they ceased when the owners of Ansett took over East West. In 1987, the system looked as immune from competition as it ever had.

Public policy changes

Deregulation may also be explained in terms of 'public interest' factors. It did not take place in a political vacuum – it was undertaken by a government which had explicitly committed itself to reform in a number of areas. Telecommunications has been partially deregulated, the financial sector has been extensively deregulated, industry protection has been reduced, and some reform of wheat, shipping and waterfront regulation is promised. Many of these changes have been qualified or been partial, and the government has shown itself unwilling to tackle vested interests too directly – it has preferred a 'consensus' approach to reform (see Keating & Dixon, 1989). This fits in well with its willingness to deregulate the airline industry to an extent that did not offend the major groups in it too much.

Another relevant factor has been the growing perception of the costs of the Two Airline Policy. A decade ago, most public criticisms of it centred on the absurdities of parallel scheduling. Since then, there have been a succession of studies, both public reports and academic papers, which have subjected the industry to scrutiny, and generally concluded that there are significant inefficiencies. The results of deregulation overseas, especially in the US, have become available, and it has become apparent that not only is a deregulated industry a feasible option, but that also, it can bring significant gains. In short, the costs of the Policy are much better understood than they were a decade ago. This has made the government more willing than before to contemplate deregulation – possibly for the 'public interest' reason that it wishes to have an efficient industry, and possibly for the 'private interest' reason that it sees the vote catching potential of lower air fares for travellers being achieved at little cost to other influential groups.

3.6 The scope of deregulation

The policy changes

The policy statement by the Minister for Transport and Communications on 7 October 1987, announcing domestic airline deregulation, was the culmination of a gradual move towards deregulation that had been taking place throughout the 1980s. It was a major break with what had gone before. The Two Airline Policy had been altered at previous times, notably in 1981, but it had been essentially kept intact. The 1987 statement promised abolition of the Policy, and the removal of most of the direct regulation of domestic airlines.

The 1981 legislation set out a timetable for changes to or abolition of the Policy. The Policy was to run for five years, from June 1982 to June 1987. After this, the government could terminate the Policy, but if it did so, it was required to give three years' notice. In June 1987, the Minister announced that the arrangements as they existed would be terminated in 1990, though he did not specify what arrangements would replace them. The October 1987 statement gave details of the policy to be implemented.

The changes in detail

It is important to recognize from the outset that the changes proposed for the domestic airline industry are substantial in their content, but limited in their coverage. The inter-state airline industry is to be largely deregulated, except that only certain airlines are to be permitted to enter it. In the main, international airlines, even including Qantas, are to be excluded from the domestic routes (though Qantas is to be given limited rights). Airlines which are permitted to operate will be subjected to very little regulation.

Such deregulation involves a considerable number of distinct changes to the various regulations, such as those affecting fares, capacity, safety and other aspects, and it is important to consider them in some detail. There are also some changes which may not be considered to be 'airline deregulation' as such, but which are taking place along with it, and which will directly affect how it works – the most important of these concern the infrastructure, airports and terminals. In considering the changes that make up airline deregulation, we consider first those affecting entry and capacity controls, and those affecting fares. It is entry, capacity and fares regulation that has formed the core of the Two Airline Policy.

Entry and capacity controls
As from 1990, it will be possible for new airlines to operate on inter-state routes. Any new airline may enter, as long as it is not an international airline serving Australia (see 'International airlines and the domestic market' below). Up till now, it has been doubtful whether such airlines could have done so; at any rate

the issue was resolved in the negative, since the government would have denied these airlines licences to import aircraft. From 1990, controls on the import of aircraft will be removed. It will thus be relatively straightforward to set up an airline.

Controls over capacity will also be removed. Up till now, the capacity that Ansett and Australian Airlines have been able to deploy on competitive routes has been controlled (and set equal for each airline). It will become possible for market share on competitive routes between the two major airlines to vary from the 50/50 share (approximately) enforced by the capacity determinations. Thus, a major disincentive for the airlines to compete for traffic will be removed. All inter-state routes, including those not flown at present, will be open for all airlines in the industry. Airlines other than the two majors have been permitted on to some inter-state routes, usually with conditions attached. Trunk routes (defined precisely in the 1981 legislation) were reserved for the majors – from 1990 the notion of a trunk route will have no special significance.

The result of these changes will be a domestic airline system in which entry to the industry and to routes, will be quite free. It will be comparable to the situation in the US.

Fare regulation
The Two Airline Policy created a degree of monopoly power in the industry by restricting the operators to two, but limited the fares charged through fare regulation. Along with entry controls, fare regulation is to be abolished, and airlines will be free to set whatever fares they wish. This involves termination of the Independent Air Fares Committee. Fares control has affected the maximum and minimum levels of fares charged, and also the structure of air fares. Airlines will have freedom to adopt air fare structures that maximize the efficiency with which their capacity is operated. The removal of fares control also means that there will be removal of *de facto* profits control. While the airlines have not been subjected to explicit profits control, and the different rates of profitability of Ansett and Australian have confused the issue for much of the period of fare control, the consequences for profitability have been taken into account when fares are set.

Intra-state regulation
Deregulation only affects inter-state routes directly, i.e. those routes under Commonwealth jurisdiction. These routes represent the bulk of the main traffic routes in Australia. However, there are some routes which are entirely intra-state, especially in NSW, Queensland and Western Australia which are of some consequence. Some states have already deregulated, as has South Australia, (see Starkie & Starrs, 1984) while others have a policy which allows scope for some competition on major route (WA).

Trade practices regulation
Domestic airlines have been hitherto partially exempt from Trade Practices

Legislation (several of the provisions of airline regulation have been in contradiction to provisions of it). After deregulation, they will become subject to this form of regulation, which involves controls over mergers, exclusive dealing, vertical restraints, collusion, resale price maintenance, and predatory pricing. Trade practices legislation has been applied in particular areas, e.g. the Trade Practices Commission has examined mergers, such as that between Ansett's owners and East West, to determine whether market dominance has resulted (the Commission has attempted to limit the scope of this merger as it affects intra-state NSW and Western Australia routes). No special legislation is to be introduced to control trade practices in the airline industry – it is to be subject to the general legislation like other industries.

Safety
Safety regulations are to be revised, and a Regulatory Review Panel is to be established to advise on this. There is a recognition that economic deregulation could affect safety performance, but it expected that safety regulations will be effective in maintaining safety standards.

Australian Airlines
Australian Airlines has been operated as a government commission, and this has had both advantages and disadvantages. The financial discipline on it has not been strong – it has been permitted to operate for long periods without earning a commercial rate of return, but it has lacked freedom in terms of capital raising and it has been subject to many accountability controls which have limited its flexibility. The government is revising its operating policy.

Qantas
The policy of distinguishing international and domestic aviation is to be continued, though it is to be relaxed slightly. From July 1988, Qantas has been permitted to carry international passengers of other airlines on domestic segments of international flights (e.g. Tokyo–Sydney–Melbourne passengers of Japan Airlines between Sydney and Melbourne or a Los Angeles –Sydney–Melbourne flight). It had the right prior to 1979; it was then restricted to fly only its own passengers on domestic segments.

International airlines and the domestic market
The rights to carry passengers of other airlines on domestic segments of international routes will not be extended to other international airlines serving Australia. This is similar to policy in force in most other countries. However, there is a major peculiarity, in that international airlines *not* serving Australia will be permitted to enter the domestic market.

Airways, airport and terminals
Access to the infrastructure of aviation – airways, airports and terminals, will have an important impact on the way deregulation works. There are a number

of changes that are proposed which either coincide with deregulation, or are part of the overall package. For example, incumbent airlines may be given preference over new airlines; this may make entry to particular routes difficult. Congested airways are not a major problem for Australia. Pricing policies for facilities can affect the way airlines compete (e.g. they may have an important bearing on the cost structures of commuter airlines). In the US problems of access at airports have had a significant (and negative) effect on the way the deregulated industry has performed (Kahn, 1988; Morrison & Winston, 1989).

While the airways question is not important, the airport question is. The key issues here are the allocation of scarce capacity and the pricing structure. Some airports are becoming quite congested (especially Sydney) and the capacity allocation problem is important – essentially the way it is solved determines which airlines can serve Australia's largest market.

The major Commonwealth airports are to be operated by the new Federal Airports Corporation, a public enterprise. The way it operates depends partly on the guidelines it is set, e.g. its profits targets and so forth. The guidelines that have been set provide little direction as to to how it is to handle its role as a key supplier to the aviation industry, especially in terms of its capacity rationing and pricing policies. It can adopt a broad range of policies which can have quite different implications, and it does not appear to have been given incentives to facilitate the working of the aviation policy.

Access to terminals is critical for airlines, and here also the policy is not very clearcut. Terminals at major airports are owned or leased by the airlines from the government. The October 1987 statement indicated that provision would be made of terminal facilities for new entrants. Since then, the government has renewed leases long-term for the two major airlines, and, while details of these leases are sketchy, it appears that only limited provision is to be made, and this is to be on terms dictated by the lessee airlines. This, in effect, means that access will be highly restricted, and that this will prove an early problem in the way deregulation proceeds.

The scope for further deregulation

As indicated above, the deregulation package as proposed does not go as far as it might in the direction of creating an open market for domestic aviation. Three areas of possible further change are discussed here, of them the second and third are by far the most important.

Privatisation of Australian Airlines
Changes in ownership of the airlines are not essential for deregulation to work effectively. It will be necessary for Australian Airlines to be operating independently without preferred access to government support – one way of ensuring this would be to privatise it.

Improving access to the infrastructure

Access to airports and terminals may be a problem for new airlines attempting to enter the market. There is little to suggest that this problem has been systematically addressed in the proposals so far. It is not so much a matter that further deregulation, in the sense of removal of existing regulations, would resolve this problem, but rather a matter of ensuring that the capacity owned by government enterprises is used most efficiently to facilitate competition between the airlines. As it stands, the Federal Airports Corporation is an enterprise with considerable monopoly power, and discretion over pricing policy, which has no incentive or requirement to use its facilities in a way to advance the stated aims of the government's deregulation policy. Indeed, the creation of FAC, its removal to an arm's length from the government, coupled with profits targets, may induce it to adopt financial, and pricing, policies which hinder rather than help competition in the airline industry.

The case of terminals is more difficult to resolve, granted that contracts for leases have been agreed. The evidence available suggests that these leases will restrict competition; while some provision is made for new entrants, it is done on disadvantageous terms to them. This need not have been the case, since the leases were negotiated after deregulation was announced. One option would be to renegotiate the leases so as to provide better access for new airlines, and to compensate the incumbent airlines accordingly. If the cost of such compensation is high, this suggests that the original leases were very advantageous to the incumbents, and hence quite anti-competitive. Apart from this, it may be possible to make better provision through new construction, perhaps by the FAC.

Lessening the domestic/international dichotomy

The main way in which the deregulation proposals are restrictive lies in the maintenance of the strict dichotomy between domestic and international markets; more thorough deregulation would involve breaking this down. Various stages can be distinguished. That which involves least change would be one in which Qantas were permitted to operate domestic flights. A more extensive proposal would be to form an Australia–NZ, or South Pacific, airline market. Beyond this it would be possible to allow participation in the Australian domestic market by international airlines which fly to Australia. These three extensions to deregulation are considered in turn.

Allowing Qantas domestic rights has two main advantages. It would add the largest Australian-based airline to the group competing in the domestic market, and it would enable Qantas to integrate domestic and international flights as it wishes, so that it can offer a more attractive overall network to home and overseas travellers. This would enhance its ability to compete in the international marketplace. Doing this would involve Australia in no further negotiations with other countries. The reason given in the October 1987 statement for not doing this is not a strong one. It is claimed that it would result in pressure for the domestic airlines to be allowed to fly on international routes.

This is probably true, but there is no necessity to award these airlines such rights (if they can be negotiated with other countries) simply because their competitive position is made tougher (surely that is the purpose of deregulation).

A more extensive form of deregulation would involve creation of an Australia–New Zealand aviation market, or a South Pacific aviation market, which would involve countries of the South Pacific other than Australia and NZ. This grouping would form an open market, which would be as open as the Australian domestic market, or subject to an agreed set of regulations. Countries of the European Community are moving towards such an integrated market in Europe. Between points in the region, aviation would be free or subject to the same regulation. Thus, Australian Airlines would be able to operate between Canberra and Wellington, and Wellington and Christchurch, while Air New Zealand could fly between Canberra and Brisbane.

The advantages of this wider market are at least twofold. National boundaries are artificial constraints so far as airline networks are concerned. With freedom to fly anywhere in the region, the airlines would be able to develop the network which is most efficient for their traffic; this will lead to operational economies, and simpler journeys for some travellers. Rearrangement of networks was a major consequence of US deregulation, and a major source of gain. The other advantage would be a market with more competitors. Air New Zealand could prove to be an effective entrant in the Australian market.

The 7 October statement rules out such a development, though it does not present any coherent reasons why. It claims that it would require granting domestic airlines 'full international rights', and that the Australia–NZ market is too small for multiple designation of international routes – a surprising claim granted that multiple designation already exists on many routes (Air New Zealand and Qantas both fly them). If anything, creation of the wider market would give opportunities for rationalisation on those routes which are already served by two airlines. The suggestion that formation of an airline consisting of all major airlines of Australia and NZ would be necessary is extraordinary.

Formation of a South Pacific aviation grouping would not mean that it could be treated as a single country for international purposes. International agreements would be unchanged, and for example, Qantas could not fly Auckland–Tokyo and Air New Zealand could not fly Cairns–Tokyo. Over time, with agreement between the countries in the grouping, such rights could be negotiated with other countries. The main purpose of formation of a wide market would be to allow greater flexibility in the internal market; this can be achieved by agreement between the countries involved, and it would be the main source of gain. It would not be necessary for the countries to be treated as one country for international negotiation purposes, though there would be gains from doing this.

A final extension of deregulation would be to open up the domestic to international airlines. This could involve rights to pick up passengers on domestic segments of international flights, or rights to schedule purely domestic flights. Allowing open access to international airlines is rare, though some countries allow international airlines to pick up domestic passengers in certain circumstances. Allowing international competition would have the same advantages as forming a regional grouping – it would enable more efficient networks, and it could increase competition. There may be questions of who the gainers would be, e.g. the gains from improved networks might not accrue to Australia.

3.7 The performance of the deregulated industry

In this section, the theory and analysis of experience of other deregulated industries will be used to suggest how a deregulated Australian domestic airline industry might perform. Naturally, it is not possible, given the uncertainties present, to make definite projections, however it is possible to indicate the range of likely possibilities, and to indicate what factors will be critical in determining the outcome.

How many airlines can Australia support?

The airline industry is not one characterized by extensive scale economies or precise minimum efficient scales, so that it is not possible to determine industry numbers from cost conditions. The practical limits come from the importance of networks, density and frequency on individual routes, and the variety of differentiated products (and hence, the possibility of market niches). (On the importance of these factors in the US, see Caves *et al.*, 1984; Call & Keeler, 1985; Morrison & Winston, 1989.) Airlines do not have to be big to survive – small airlines can operate efficiently. Rather, they have to have markets to survive – they can gain markets by having networks, regional dominance, or market niches. To determine how many airlines can survive, it is essential to look at the number of independent firms that can operate on individual routes.

There seems to be a practical limit to the number of airlines that compete on a route – this limit depends partly on route density. Suppose, between two cities, there is enough traffic to support 40 flights per week. Technically it would be possible for 40 airlines to offer one flight per week. There are, however, scale economies in route operation – two flights per day are not twice as costly as one flight per day. Further, there are strong advantages to being able to offer passengers a convenient schedule – the 40 flights by independent airlines would be extremely inconvenient for travellers. The scale economies probably peter out, as do the frequency advantages, and they are matched by advantages of choice (networks, service standards, reputations) gained from having additional airlines. Thus, routes of particular density and length tend to support particular numbers of competitors.

This is seen in Table 3.11 showing the effective numbers of competitors in US

Table 3.11 Effective numbers of firms serving city pair markets. (Weighted averages by year, distance, and passenger density).

Year	200–500	501–1,000	Miles between cities 1,001–1,500	1,501–2,000	2,001+
			25–50 passengers per day		
1983[a]	1.24	1.57	1.93	2.37	1.82
1987[b]	1.45	1.88	2.23	2.70	2.52
			51–200 passengers per day		
1983[a]	1.43	1.89	2.22	2.27	2.17
1987[b]	1.36	2.04	2.56	2.90	2.84
			201–500 passengers per day		
1983[a]	1.50	2.25	2.46	2.30	2.46
1987[b]	1.61	2.06	2.52	2.72	2.94
			501–1,000 passengers per day		
1983[a]	1.90	2.25	2.43	2.45	2.88
1987[b]	1.90	2.28	2.37	2.18	3.82
			Over 1,000 passengers per day		
1983[a]	2.33	2.80	2.67	2.83	3.85
1987[b]	2.22	2.92	2.45	2.83	4.13
			All densities		
1983[a]	1.81	2.15	2.43	2.42	2.72
1987[b]	1.80	2.23	2.46	2.67	3.27
			Average for all markets		
		1983	2.40		
		1987	2.49		

Source: Taken from Congressional Budget Office (1988): 17.
Notes: (a) First quarter
(b) First quarter, after adjusting for mergers taking place after the first quarter of 1987.

city pair markets in 1983 and 1987. These are averages, and they mask greater variability between individual markets. They are notably stable: in 1983 there were many more firms in the industry as a whole than in 1987, after the mergers and exits. As expected, the number rises with market density, but by no means in proportion to it, even quite small routes can sometimes support a couple of firms. This gives us an idea of the level of competition sustainable in route markets; in the US, there are several airlines and free entry on to routes, and if more airlines wished to operate on individual routes, they undoubtedly could.

This information can be used to get an indication of the number of airlines that might be able to compete on main Australian routes. The top 20 routes (accounting for 77% of total passengers), their distance and traffic are shown in Table 3.12, along with the number of competitors that they might support (derived from 1987 values, Table 3.11). Australia, reflecting its concentrated urban pattern, has a group of routes which, by international standards, are dense, and a much larger fringe of low density routes.

This Table can give an indication of the number of airlines that might survive in the whole domestic market. In the US, no airline serves a majority of the routes that exist. The major airlines serve the major cities, and have wide coverage of the nation by extensive use of hubbing (serving A to B indirectly

Table 3.12   Effective numbers of firms on Australian routes.

Route numbers 1		Distance (miles) 2	Passengers/day** 3	Firms 4
1	Melbourne–Sydney	439	6,679	2.22
2	Brisbane–Sydney	465	4,372	2.22
3	Adelaide–Melbourne	404	2,184	2.22
4	Coolangatta–Sydney	422	1,809	2.22
5	Canberra–Sydney*	147	1,728	2.22
6	Brisbane–Melbourne	857	1,332	2.92
7	Adelaide–Sydney	724	1,326	2.92
8	Canberra–Melbourne	292	1,170	2.22
9	Hobart–Melbourne	379	1,081	2.22
10	Melbourne–Perth	1,683	1,079	2.83
11	Brisbane–Cairns	865	957	2.28
12	Launceston–Melbourne	291	824	1.90
13	Melbourne–Coolangatta	828	784	2.28
14	Perth–Sydney	2,037	755	3.82
15	Brisbane–Townsville	692	676	2.28
16	Adelaide–Perth	1,316	666	2.37
17	Brisbane–Rockhampton	323	416	1.61
18	Devonport–Melbourne	251	357	1.61
19	Karatha–Perth	777	345	2.06
20	Brisbane–Mackay	497	344	1.61

*   Calculated on the basis of Table 3.11.
** Year ending June 1988.
   Sources: Columns 1, 2 and 3 from Department of Transport and Communications, *Domestic Aviation Statistics*.
   Column 4 from Table 3.11.

through their hub at C). While some hubbing is likely in Australia, the scope for it would not be as great (fewer centres of population) and an airline, to have an effective network would need to serve most of the main (say 10) routes. This would give a practical limit of about three independent airlines (this would result in the smaller routes having more competitors than the *average* for comparable US routes). It would be possible for there to be more moderately large airlines if they were regionally based – for example if a Queensland-based airline had services to Melbourne and a Victoria/NSW-based airline had services to Queensland. However, if this were the case, there would be a tendency for such airlines to merge (as they have in the US) to gain a better coverage of the nation. Over the longer term, there would be a practical limit of about three airlines with networks covering the major routes.

Beyond this, there may be a fringe. Apart from the commuter airlines (which are likely to be linked commercially to the major airlines), there would be a small group of regional airlines with one two inter-state routes. Another possibility is that of small airlines that have managed to find niches (e.g. in tourism markets) that the major airlines are not interested in serving. Finally, there is the possibility of competition from international airlines.

One possibility is that Qantas or Air New Zealand might operate some domestic routes on a restricted basis – in such a case, they would form part of the fringe. If either or both were permitted to operate freely, and they chose to do so, they would inevitably form part of the small group of major airlines – possibly forcing an incumbent out. Their international links would give them advantages that no domestic new entrant could match. Allowing, say, Qantas to operate domestic routes would not increase the number of major airlines by one – in the long run it would probably leave numbers constant. Finally, there is the possibility of allowing other international airlines to operate on a restricted or open basis on domestic routes. They are likely to do so only to a limited extent, operating those routes which tie in well with their international services. Thus, a small handful of routes could have 4–6 airlines competing, but of these, perhaps two would have a limited involvement, perhaps one flight per day, run in conjunction with their international flights. While unlikely to be of quantitative importance, such airlines could affect the competitive outcome in markets they serve.

Entry barriers

The ability of new firms to enter and compete may be important in determining how well an industry performs. If there is a group actively competing and efficiently operating, entry conditions need not be important. If there are few firms competing, if collusion is present, and if they are operating inefficiently, entry conditions will be important. If entry is easy, new firms can enter and actually compete, and force efficiency improvements. Even if entry does not actually take place, potential entry may lower prices, and, possibly more importantly, will induce incumbents to produce efficiently, so that they can survive if entry does take place.

Cost advantages and disadvantages
One of the traditional barriers to entry comes from cost advantages held by the incumbent over the entrant. If this barrier is present, the incumbent can produce more cheaply than the entrant, who will find it difficult to compete. This is not likely to be the case if the entrant has equally good access to inputs (see 'Access to essential inputs' below). The technology is the same for entrant and incumbent, and they have equal access to it. In the US, the entrants had not built up, over periods of regulation, inefficiencies which took time to remove, nor were they tied to expensive labour contracts. They were thus able to operate at lower cost, for the years immediately after entry, than the established airlines (Bailey *et al.*, 1985, ch. 5). This is only a temporary advantage. The evidence on cost levels of Australia's airlines suggests that they are not as low as would be feasible, and that a new entrant might be able to operate at lower cost, thus giving it a temporary advantage over the incumbents.

Networks and product differentiation
The analysis of barriers to entry suggests that economies of scale need not be a barrier (since scale economies in production are equally available to entrants

and incumbents) but that combined with a product differentiation or marketing advantage, they can constitute a barrier (the entrant cannot sell as much as the incumbents, and must operate at higher cost per passenger). The same is true with network effects. An entrant can match or surpass the network of the incumbents, at least on a technical level, and offer as attractive frequencies, however it will have great difficulty in filling its aircraft. It may have to enter with a more restricted network, and this gives it a disadvantage *vis-à-vis* the incumbents. A new entrant in the Australian market may have considerable difficulty in selling its services.

Access to travel agents is a determinant of market share. If an airline has links to, or owns, a travel company, it can use it to direct traffic its way. Many of the larger travel firms in Australia already are partly owned, or have strong links, to established airlines. This will make entry more difficult (except for Qantas and Air New Zealand). A similar advantage will be ownership of, or preferential treatment in, a computer reservations system. The two major domestic airlines possess their own, early type, CRSs, and are moving towards participation in the British Airways-United Appollo system. Qantas is linked to American Airlines' Sabre system. Here the advantage lies with the first mover, and the incumbents have moved first (unless the Sabre CRS becomes dominant in Australia). This is not an absolute advantage; other airlines can be linked to the system, though they will be given lower priority in displays, and may have to pay more for access. As a potential entrant, Qantas may have an advantage over the domestic airlines, and Air New Zealand would be no worse placed than them. Other entrants will be at a disadvantage however. For some small airlines, it may be possible to tap markets that do not rely heavily on CRS facilities, for example, all inclusive holidays.

Access to essential inputs

If the entrant does not have access to essential inputs on the same terms as the incumbents, it will be at a disadvantage, and may not be able to compete at all. In the short to medium term, this may be the most serious entry barrier in Australia.

Access to airports is obviously critical. There is no reason to expect this to be a problem unless owners of airports, such as the Federal Airports Corporation, or local authorities, give preference to existing users (which they well might). With crowded airports such as Sydney, preferential access policies could result in a strong barrier to entry. It is possible to make airport facilities (apart from terminals) equally available to all, at an appropriate price. This is an area where choice of policy can have a major effect on the outcome of deregulation.

Terminal facilities are likely to provide the most difficult access problem in the short to medium term. In the main, the two incumbent airlines have long leases for their terminals at major city airports, and while they are required to make some space available to other airlines, this will be done on their own terms, and on a restricted basis. Few airports have much available additional capacity which the operator can allocate to new entrants. There is scope to use international terminals for domestic flights, particularly if the airline using them is already an international airline (e.g. Qantas). Some international terminals, especially that of Sydney, are already crowded, though some expansion is taking place.

Performance under deregulation

The outcome of deregulation will depend on how competitive the industry is, and how easy it is to enter the industry. It is not possible to forecast these accurately, as they depend on uncertain reactions and on policy decisions yet to be made; further, experts will disagree as to how competitive the industry will be. Various scenarios can be envisaged: (a) a monopolistic industry, or one with strong dominance by a few airlines, which may be tacitly colluding; (b) a competitive industry, which, although not characterized by large numbers of firms, has active competition between firms and easy entry; (c) an industry which is open to international competition, and in which there is some participation by international airlines, which do not dominate the industry but put pressure on other airlines to be competitive. Scenario (c) is not feasible under current proposals. It would be the most open and competitive of the three suggested here. Other scenarios are possible, for example, a few overseas international airlines, if permitted to enter, might dominate the industry and earn profits at Australia's expense. This is an unattractive scenario, and it can be prevented by policy decisions.

We consider that the most likely outcome is an industry which will fall between scenarios (a) and (b), assuming international airlines are excluded, as proposed. If they were permitted to enter, the industry would be more competitive. The degree of competition is not likely to remain constant over time – there will be phases when the industry will move towards scenario (a) and then towards scenario (b).

Deregulation can have an impact on the different aspects of efficiency in several ways. Removal of *de facto* profits control can increase the incentive given to firms to minimize costs and seek out markets – it can also increase the incentive to set prices above costs. Additional competition, potential or actual, can force firms to minimize costs, and actual competition will force firms to keep prices closer to costs. Deregulation *could* mean that the two major airlines compete more aggressively between themselves, even if new entry does not, and is not likely to, occur. The main problems with regulation were that firms did not need to minimize costs, or service all markets. Whether or not there is more competition for the major airlines, deregulation will provide them with more incentive to, and additional pressure to, minimize costs and serve all markets. Even if a monopoly, or co-operative duopoly is the result, it is likely that there will be net gains. The gains from lower costs and all markets being served would not be cancelled out by the efficiency losses from more monopolistic pricing.

The gains from deregulation will be greater if international airlines, especially the regionally-based Qantas and Air New Zealand, are permitted to operate domestic flights. They are likely to increase competition, since they are the strongest potential entrants, though they may not increase absolute numbers in the industry – the entry of Qantas might dissuade other potential entrants from trying. Furthermore, they are likely to improve networks – it makes sense for some airlines flying to/from Australia to be able to offer an integrated network of domestic and international flights.

The main gain from deregulation will be from increased pressure for firms to produce efficiently. It is perhaps optimistic to suggest that deregulation will lead to productivity levels equal to the best achieved overseas; the Australian market will not be as competitive as the US market, and economy-wide labour market practices differ. However, a gain, achieved over a period of a few years, of about 10% would produce major savings to the economy. If cost differences are of the order of 15–20%, this should be quite achievable.

Gains from better fare structures and increased availability of low fares would also be significant, though they are unlikely to amount to as much as the productive efficiency gains. Average fares and yields will fall, but these will exaggerate the gains, since the convenience of service, on average, will decline; the gain comes from offering passengers the option of lower fares at the expense of less convenience.

Network and scheduling effects are likely to be much less important. Under the current proposals, there would not be any major network effects, though if the distribution between domestic and international aviation was weakened or removed, it would be possible for the carriers based in the region to devise more efficient route patterns. There is no certainty that parallel scheduling would disappear, but as the major airlines become more different, and as smaller operators seek out niches not well served, the incidence of parallel schedules will lessen.

There will be a range of other effects, some of them desirable and others undesirable. These may be of considerable importance for some individuals or firms, but overall they are likely to be of much less consequence than the first two effects. Many are such that they represent gains to some and losses to others. The main source of inefficiency loss is likely to come from excessive demands on the infrastructure which deregulation could (but may not) bring. These losses will be temporary (they can be eliminated by additional capacity) and they can be minimized by policies to ration scarce capacity efficiently.

3.8 Assessment

Australia's domestic airline deregulation is partial in the sense that the separation of international and domestic airline markets will be preserved.

Within Australia, regulation will be more or less completely removed, to leave a situation where the two major domestic airlines have very considerable advantages, especially in their access to essential inputs, such as terminals. New entry will not be restrained by regulation, but it will be difficult, and most of the natural potential entrants, existing airlines operating in the region, will be prevented from entering.

It seems likely that the two major airlines which have grown up under the mantle of the Two Airlines Policy will continue to be the largest, and probably dominant, firms in the industry. One or two successful entrants of any size is the most that can be expected, and it is possible that no entrant will succeed. The industry may appear to have been little changed by deregulation. Appearances could be deceptive, however.

The major airlines will be facing change on at least two fronts. They may not face more actual competition, but will certainly face the threat of competition. Airlines do not seem to be contestable as was first thought, and potential competition is not as strong a market discipline as was first thought. This is especially true in respect of pricing behaviour – firms can lower prices rapidly if entrants appear, and thus they do not need to moderate their pricing behaviour all the time. Potential competition is possibly more effective as a discipline to firms to keep costs down and produce efficiently – if new firms appear with lower costs, they may not have enough time to respond if their costs are too high. The strength of this discipline depends on how difficult it is to enter the industry.

The second front on which change will come is in terms of the incentives and constraints faced by the firms. It will no longer be the case that profits are (informally) constrained. A private Ansett, and a more commercially-oriented government-owned Australian Airlines will be able to earn and keep profits. They will have a stronger incentive than they did while operating under regulation to keep costs low, and serve any market which promises a profit.

The worst scenario would be one where entry was difficult, there were few efficiency gains to be made, and the two airlines effectively colluded to raise fares to a profit maximizing level. However, even if market power exists, and it is used at a cost in terms of efficiency, it is likely that other efficiency gains could be made, and that these would outweigh the losses from monopoly pricing. Profit-oriented airlines would have an incentive to reduce costs and serve all markets, even though they will appropriate most of the benefits from doing so to themselves. A monopolistic industry is likely to perform better than an industry regulated in such a way as to weaken incentives for efficiency.

It is possible that the two airlines will compete between themselves. This will lessen the scope for raising prices above costs, and increase the pressure to minimize costs and seek out all markets. There is some possibility of effficiency costs, since prices need not be kept at costs, as they have been under regulation. These costs are not likely to be large relative to the gains from the other sources identified. Prices may fall, even though they may be a higher proportion of costs.

The best results would occur if entry is easy, and successful entry does occasionally occur, and there are more than two operators on a range of routes. Entrants can be relied upon to seek out any markets worth serving, and the pressure to keep costs at a minimum will be greatest.

The actual situation is likely to fall between the second and third of these possibilities. Entry seems feasible, though not easy, and some firms are planning to enter the market. The major airlines are responding to the possibility of changed circumstances following deregulation. They are attempting to reduce costs through more efficient use of manpower, and they are showing greater interest in serving low-fare markets. Even if entry does not take place, or it is unsuccessful, the possibility of it occurring will have had positive effects.

In terms of outward appearances, especially in terms of airline numbers, domestic airline deregulation may not produce major changes. In terms of airline behaviour and performance, it may. The overall efficiency of the industry is likely to increase, though there will be some aspects of efficiency (relationship of price to cost) which could deteriorate.

Once deregulation has occurred, there will still be a role for policy as a determinant of performance. The industry would be more competitive, and have the scope for better networks, if it were opened to international competition, especially from airlines based in the region. As US experience shows, access to airports and terminals is an important determinant of how effective competition is. This will be a major problem in Australia, and it remains to be seen whether facilities in heavy demand are rationed efficiently, or used to strengthen the market dominance of the incumbent airlines. There is also the question of whether restrictive practices legislation can be applied effectively to lessen the anti-competitive conduct of the airlines. The success of deregulation, measured in terms of the gains achieved, will depend to a considerable degree on the response of the government to these problems.

References

Bailey, E.E., & Williams, J.R (1988) 'Sources of economic rent in the deregulated airline industry', *Journal of Law and Economics*, April, 173–202.

Bailey, E.E., Graham, D.R. & Kaplan, D.P. (1985) *Deregulating the airlines* Cambridge, Mass: MIT Press.

Blain, N. (1985) *Industrial relations in the air*, St Lucia: University of Queensland Press.

Brogden, S. (1968) *Australia's Two Airline Policy* Melbourne: Melbourne University Press.

BTE (Bureau of Transport Economics) (1985) 'Competition and regulation in domestic aviation: submission to independent review. *Occasional paper No. 72.* Canberra: Australian Government Publishing Service.

Call, G. D. & Keeler, T. E. (1985) 'Airline deregulation, fares and market behavior: some empirical evidence', In A. F. Daughety, *Analytical studies in transport economics*, Cambridge, England: Cambridge University Press.

Caves, D. W., Christensen, L. R. & Tretheway, M. W.(1984) 'Economics of density versus economics of scale: why trunk and local service airline costs differ', *Rand Journal of Economics* **15**(4) 471–89.

Centre for Independent Studies (1984) *Changes in the air? issues in domestic aviation policy*, CIS Policy Forums, 3, Sydney.

Congressional Budget Office (1988) *Policies for the deregulated airline industry*, Washington.

Corbett, D. (1965) *Politics and the airlines*, London: Allen & Unwin.

Department of Transport (Australia) (1979) *Domestic air transport policy review, report and appendices*, Canberra: Australian Government Publishing Service.

Forsyth, P. J. (1985) 'Can international comparisons of air fares be made?' Discussion Paper 133. Centre for Economic Policy Research, Australian National University.

Forsyth, P. J. & Hocking, R. D. (1980) 'Economic efficiency and the regulation of air transport' Monograph M62. Committee for Economic Development of Australia.

Gallagher, F. D. (1980) *Study of air transport policy for the Northern Territory*, Darwin: Northern Territory Department of Transport and Works.

Gannon, C. A. (1979) 'Parallel scheduling', Appendix A.12, Department of Transport, *Domestic air transport policy review*, Appendices, Canberra: Australian Government Publishing Service.

Harper, I. (1985) 'Why financial deregulation?', Discussion Paper No. 132, Centre for Economic Policy Research, Australian National University.

Hocking, D. M., & Haddon-Cave, C. P. (1951), *Air transport in Australia*, Sydney: Angus & Robertson.

Hocking, R. D. (1972) 'Some economic aspects of Australia's Two Airline Policy', Monograph M35, Melbourne: Committee for the Economic Development of Australia.

Holcroft, W. (1981) *Domestic air fares: report of the independent public inquiry*, Canberra: Australian Government Publishing Service.

Hotelling, H. (1929) 'Stability in competition', *Economic Journal* **39**, 41–57.

Kahn, A. E. (1988), 'I would do it again', *Regulation* No. 2.

Keating, M. & Dixon, G. (1989) *Making economic policy in Australia 1983–1988*, Melbourne: Longman Cheshire.

Kirby, M. G. (1981) *Domestic airline regulation: the Australian debate*, Sydney: Centre for Independent Studies.

Kirby, M. G. (1984), 'Airline economies of scale and Australian domestic air transport policy' Discussion Paper No. 112, Canberra: Centre for Economic Policy Research, ANU.

Mackay, K. R. (1979) 'A comparison of the relative efficiency of Australian domestic airlines and foreign airlines', Appendix A6.1 in Department of Transport, *Domestic air transport policy review*, vol. 2.

May, T. E. (1986) *Independent review of economic regulation of domestic aviation*, Canberra: Australian Government Publishing Service.

Morrison, S. & Winston, C. (1989) 'Enhancing the performance of the deregulated air transportation system', *Brookings Papers on Economic Activity: Microeconomics*, pp. 61–112.

Poulton, H. W. & Richardson, J. E. (1968) 'Australia's Two Airline Policy: law and the layman', *Federal Law Review*, June.

Starkie, D. & Starrs, M. (1984) 'Contestability and sustainability in regional airline markets', *Economic Record*, Sept. 274–83.

Riley, J. G. (Chairman) (1986) *Review of New South Wales air services: flying towards 2000*, NSW: Government Printer.

WA (Western Australia, Commission of Transport and Director General of Transport) (1982), *Review of internal air services and policy*, Perth.

Wettenhall, R. L. (1962) 'Australia's Two Airline System under review', *Australian Quarterly*, March.

CHAPTER 4

Aviation policy in Europe*

Kenneth Button and Dennis Swann

4.1 Introduction

The 1980s witnessed considerable changes in attitude towards economic regulation. Changes which transcended international borders and which covered virtually all aspects of economic activity. While there were many variations in degree and form, European economies, both East and West, were subject to some of the most radical changes. The economies of many Eastern bloc countries had already experienced a degree of liberalization in the 1970s. With the advent of *Perestroika* the economies of many Eastern bloc countries this process was extended further with free market forces given greater power in determining production and consumption priorities in the USSR. In Western economies the decade was characterized by a withdrawal of the state as privatisation and regulatory reforms were carried through (Swann, 1988a). The movement towards the single European Market within the European Communities (EC) by 1992 is perhaps the most obvious manifestation of this trend (Commission of the European Communities, 1985) but important changes have also occurred in the domestic policies of many European nations.

Aviation, both domestic and international, is an industry which has traditionally been regulated throughout the world. Indeed, writing in the late 1960s, Lissitzyn (1968) contended that, 'probably no other world-wide economic activity of comparable magnitude is more thoroughly regulated, less free of official restraint and guidance, than is world air transport'. In the late 1970s and the 1980s, however, it was the subject of some of the most important and dramatic changes in policy. From a very highly regulated industry it has gradually become more market oriented as both national and international markets have been liberalized. The nature of such changes varies across countries. Changes in some countries, e.g. the USA (see, Levine, 1987; Keeler, 1990) have been relatively rapid and involved the major restructuring of the legal framework of policy. Equally, some countries have chosen liberalization through *de facto* implementation of existing regulatory systems with only minor legal reforms, e.g. as with domestic aviation in the United Kingdom after

*The authors would like to thank the British Academy for providing financial support for the project from which this chapter is drawn.

85

the 1980 and 1982 Civil Aviation Acts (see, Civil Aviation Authority, 1984b, 1987).

The objective of this chapter is to focus on the changing situation within Europe. Particular, but not exclusive, attention is paid to the aviation policy of the EC. Since this embraces both national and international considerations, with the need for lengthy negotiations which this inevitably entails, the nature of change has by necessity been rather less rapid and, at the same time, more messy than that experienced within some of the other major international aviation markets. The exposition form adopted is to initially provide some details of the nature of the European aviation market together with a brief discussion of some of the difficulties involved in analysing the reforms which have been occurring. This is combined with an account of policies prior to 1980, what one might consider as the period of 'High Regulation'. To understand the pressures for change and the forms that they are taking it is initially important to appreciate why market regulation was thought important in the first place.

This discussion leads on, in section 4.3, to a consideration of why the mid-1980s should have proved something of a watershed in terms of European aviation policy. One can obviously point to general trends towards liberalization of markets at the global level, but there are more detailed issues of why Europoean aviation has tended to follow these trends and what specific forces were acting to cause the timing and nature of the reforms which are taking place. This is followed by a review of some of the constraints which have both inhibited the speed of change and been influential in shaping its form. Section 4.5 provides an examination of what exactly has taken place within the EC since the mid-1980s up until the end of the decade. Since much of the reform within Europe must take place within the context of somewhat wider EC policies there is also the need to spend some time considering what forms of market control are likely to remain over the activities of the aviation sector. This is done in Section 4.6. The specific question of the implications of the emerging European situation in the wider context of the global, international aviation market is considered in the penultimate section. Finally, there are a few brief concluding observations on the developments to date and possible trends in the future.

4.2 Aviation in Europe

The actual aviation industry which has emerged in Europe in the late 1980s is both a substantial one and a diverse one. There are, for example, over 130 European-based airlines of various sizes operating in Europe. In global terms, however, European airlines are relatively small. Carriers such as British Airways, Iberia, Air France and KLM rank in the top 10 airlines in terms of international traffic but because of the relatively small size of their domestic operations they are dwarfed by the major US domestic airlines. The recent merging of British Airways (which at the time was doing 46.3 billion scheduled passenger-kilometres) and British Caledonian (doing 8.8 billion) created the largest European airline in terms of *passenger-kilometres* done but this needs to be set in the context of US airlines such as United Airlines which was then doing 106.7 billion passenger-kilometres and American Airlines doing 91.3 billion. Table 4.1 provides more up-to-date, 1988 data, on the world's 20 largest, non-Eastern bloc, carriers (by revenue). One should also perhaps note that Aeroflot did some 213.3 billion passanger-kilometres in 1988 while CAAC (of the People's Republic of China) did 25.4 billion.

Table 4.1 The world's largest airline companies 1988 (according to revenue earned).

Carrier	Country	Revenue ($ million)	Passenger-km (billion)
UAL	USA	9,014.6	111.2
AMR	USA	8,824.3	104.3
Texas Air	USA	8,572.9	111.0
JAL	Japan	7,249.7	45.5
Delta	USA	6,915.4	79.0
Lufthansa	West Germany	6,739.9	42.5
British Airways	UK	6,690.4	49.1
Air France	France	5,953.7	34.4
USAir Group	USA	5,707.0	48.9
NWA	USA	5,650.4	64.5
All Nipon	Japan	4,427.4	23.5
SAS	Sweden	4,412.0	14.0
TWA	USA	4,361.1	56.0
Pan Am	USA	4,165.1	46.7
Hanjin Group	South Korea	3,731.9	15.1
Alitalia	Italy	3,258.1	18.5
Swissair	Switzerland	2,927.8	14.8
Iberia	Spain	2,874.2	20.9
KLM	Netherlands	2,823.2	23.3
Air Canada	Canada	277.8	25.1

Source: Woods, W., 'Revolution in the air', *Fortune* **121**, 38–9.

For convenience the types of airlines now provided in Europe are often divided into a number of broad categories (Organization for Economic Co-operation and Development, 1988).

● There are the 22 'flag carriers' such as Air France, British Airways, KLM, Lufthansa, etc which provide the main intra-European and international services. These are often publically owned.
● There are carriers such as Air Charter in France; British Airtours in the UK; Condor in West Germany, etc which provide non-scheduled services. These smaller airlines are often associated with the flag carriers which frequently have financial interests in them. (See Table 4.2 for details of the linkages in 1988.)

Table 4.2   The ownership of major European Communities' airlines in 1988.

Airline	Stake in company (%)	Participation in other airlines (%)
Aer Lingus	government (100)	Aer Turas Teoranta (maj.)
Air France	government (100)	Air Charter (80)
		Air Inter (36)
		Air Guadeloupe (45)
		Euskal Air–via Air Charter (29)
Alitalia	government (67)	
	private (33)	
British Airways	private (100)	British Caledonian (100)
		British Airtours (100)
		Cal Air International (100)
Iberia	government (100)	Aviaco (67)
KLM	government (36.9)	Martinair (25)
	private (63.1)	Transavia (40)
		NLM Cityhopper (100)
		Netherlines (100)
		Air UK (14.9)
Lufthansa	government (74.31)	Condor (100)
	public institutions (7.85)	DLT (40)
	private (17.84)	Cargolux (24.5)
Luxair	government (100)	Luxair Commuter SA (100)
		Cargolux (33)
Olympic AW	government (100)	
Sabena	government (54.72)	Sobeliar (71.08)
	private (45.28)	
SAS	government (50)*	Linjeflyg (50)
	private (50)	Greenlandair (25)
		Wideroe (22)
		Scanair (maj.)
TAP Air Portugal	government (100)	Air Atlantis (100)

* Including that of other Scandinavian governments.
Source: International Foundation of Airline Passenger Associations, (1988), *European Airline Mergers*, (Geneva, International Foundation of Airline Passenger Associations).

● There are other large carriers such as Air UK, Britannia and Air Europe in the UK and Air Inter in France which serve non-schedule and regional markets. Again links with national carriers exist with these airlines providing feeder services. In some cases the links may be via holding companies such as British Airways' 40% holding in Plimsol Line which is the owner of UK regionals, Birmingham European Airways and Bryman. Sometimes these links transcend national borders, e.g. as with KLM and Air UK. (See again Table 4.2 for major linkages in 1988.) From outside the EC, Swissair has taken a 41% share in Crossair and 7% in Austrian Airlines.
● At a lower level there are some 60 or so small airlines (with less than 250 employees) which conduct local, charter and minor cargo operations especially on thin routes and to the more remote regions.

The actual demand for air travel has grown considerably in Europe over recent years and has outstripped most of the predictions which have been made.

For instance, in the financial year 1987 to 1988 alone, the growth at European airports was some 10% compared with a forecast of 5–6%. Table 4.3 provides an indication of the growth in both aircraft kilometres and passenger kilometres provided by the airlines of the main European nations in the decade from 1976. Further, the current forecasts are that this growth will continue with European traffic doubling by the end of the century.

In aggregate terms, if one just considers EC airlines providing scheduled services, then towards the end of the 1980s (see Table 4.4) they accounted for 114.3 million passengers carried per annum and were enjoying annual operating revenues of over \$28 billion. Taking a larger area, in 1985 airlines from the 22 European Civil Aviation Commission (ECAC) states carried 120 million passengers, some 92 thousand-million passenger-kilometres on scheduled services and a further 42 million passengers, some 67 million passenger-kilometres on non-scheduled services. All this should be set in the medium period context of not only rising trends in passengers carried but also in terms of more seat-kilometres being offered and higher revenues being earned throughout the latter part of the decade. It should also be taken in the context of scheduled services only forming part of the overall market. In terms of passengers carried, for example, as we see in Table 4.3, European airlines also provide very significant numbers of non-scheduled flights. Just taking the data above, for instance: in 1985 non-scheduled services accounted for some 26% of

Table 4.3 Selected outputs of European airlines for 1976 and 1986.

| | Scheduled | | | | Non-scheduled | |
| | Aircraft km (million) | | Passenger km (billion) | | Passenger km (billion) | |
Country	1976	1986	1976	1986	1976	1986
Belgium	48	53	3.8	5.6	0.5	0.0
Denmark	37	42	2.3	3.5	0.1	1.4
France	250	290	23.0	39.5	0.4	0.1
Greece	35	49	3.4	6.4	0.0	0.2
Ireland	19	21	1.5	2.5	0.6	0.6
Italy	130	140	11.0	16.9	0.7	0.7
Luxembourg	4	4	0.1	0.1	–	–
Netherlands	95	130	10.0	19.8	1.2	2.0
Portugal	42	40	3.3	4.5	0.9	0.0
Spain	130	160	11.0	19.1	0.2	3.7
United Kingdom	290	440	28.0	65.5	17.0	25.9
West Germany	170	254	14.0	26.6	0.2	0.1
Austria	15	26	0.7	1.4	0.3	–
Czechoslovakia	26	24	1.4	1.9	0.4	0.3
Finland	30	38	1.3	2.9	1.8	2.8
Hungary	10	18	0.5	1.1	0.1	0.0
Norway	49	71	2.9	5.0	0.1	1.3
Sweden	59	91	3.6	6.8	0.1	0.0
Switzerland	83	110	7.6	13.0	0.1	0.1
Yugoslavia	29	36	2.0	3.8	–	1.7

Source: International Civil Aviation Organization (1987), *Civil Aviation Statistics for the World*, Montreal: International Civil Aviation Organization.

Table 4.4 Trends in scheduled aviation in the late-1980s.

Unit	Year	IATA carriers	% change	EC air carriers	% change
Passengers carried	1986	949.9	+6.1	97.1	+1.6
(million)	1987	1,036.8	+9.1	105.9	+9.1
	1988	1,072.1	+5.0	114.3	+7.9
Available seat kilometres	1986	2,213.0	+6.3	282.8	+6.0
(thousand million)	1987	2,367.0	+6.9	298.4	+5.5
	1988	2,506.0	+6.0	333.3	+11.7
Total operating revenue	1986	123.0	+9.6	19.9	+15.8
(thousand million US$)	1987	147.0	+19.5	24.4	+22.3
	1988	166.0	+12.9	28.4	+16.6
Load factors (%)	1986	65.2	–	64.7	–
	1987	67.2	–	69.0	–
	1988	66.9	–	68.2	–

Source: Adapted from International Air Transport Association and Association of European Airlines data.

passengers being carried, and representing some 42% of the passenger-kilometres performed, within ECAC countries were being carried on non-scheduled services (or about 60% of the total within the EC).

It is also perhaps useful to contrast air travel in Europe with that in other parts of the world. Compared, for instance, with the domestic USA market, the overall passenger-kilometres performed within Europe (ECAC members) is about one-third the number although about half as many passengers are carried. The latter reflects the shorter average flight lengths within Europe. The Japanese domestic market involves carrying about a quarter of the passengers moved in Europe but, again because of different average flight lengths, a third of the passenger kilometres. Looked at another way, the total international passenger-kilometres performed within Europe represented over 11.5% of the world's total in 1986 – this compares with the 25.5% carried on the North Atlantic route. In aggregate, however, combined intra-European international air traffic and air traffic between Europe and the rest of the world accounted for over 68% of total international passenger-kilometres. Further, as we see from Table 4.3, growth in air transport associated with the EC tended to outpace that experienced by International Air Transport Association (IATA) members as a whole during the late 1980s.

Moving away from the statistics of European aviation, one of the major problems with considering European transport is that there is at present no single market for transport services. This is as true for aviation as any other mode. First, there is the problem that aviation is itself not a single industry but can be divided into sectors providing such differentiated products as scheduled trunk services, commuter services, charter services, domestic services, inter-continental services, etc. Then there are obvious geographical factors (e.g. demographic distributions) which divide aviation markets across the Continent. But these are not the only problems nor, indeed, should they be viewed as the primary ones. The main reasons there is no genuinely single market are institutional. There are clearly national, political blocs which divide up Europe and have evolved for reasons far removed from transport efficiency. But in addition, there are a variety of important international agencies concerned with transport and related matters which have both differing memberships and functions. As a result of this, the bodies, agencies and commissions which oversee the regulation and control of the national and international aviation sub-markets which exist within Europe appear as a rather confusing tableau.

Not only are there numerous sovereign states each with its own approach to domestic and international aviation policy but there are also large economic blocs, such as the EC and Comecon, which have, to varying degrees, co-ordinated policies regarding aviation between members countries. Further, overlapping these national and community blocs are more specific aviation agencies such as the ECAC which embraces some 22 member states. There are now about 200 bilateral agreements to provide air transport services between the 22 ECAC countries. Additionally, countries are bound, in varying degrees, by wider international agreements and codes on aviation, many established at periodic 'Conferences', and conform to pricing structures set down by bodies such as the IATA - see, Doganis (1985). Also included here must be the International Civil Aviation Organization (ICAO), which is a permanent international forum of 157 member states established to oversee safety standards, and signatory countries to the International Air Services Transit Agreement (IASTA), which permits carriers from one state to transit the air space of another. Finally, while these codes and agreements have generated numerous bilateral Air Service Agreements (ASAs) - technically treaties - covering terms of operation for scheduled services between individual states, charter operations have remained separate and are covered by the 'inclusive tour charter'.

To complicate matters further, overlaying this international structure there are bodies such as the Association of European Airlines which act as lobbying groups to represent some, but not all, of the European air carriers at the various forums which exist. On the other side of the table there are groups such as the International Foundation of Airline Passenger Associations (IFAPA) representing consumers. There are also international trans-national agencies such as Eurocontrol, the Committee for European Airspace Co-ordination, the European Air Navigation Planning Group, etc, which have a range of responsibilities covering air traffic control and management matters.

4.3 The 'small step' changes of the early–1980s

Transport policy has historically and in very general terms gone through a number of distinct phases since the turn of the century. The early part of the century, broadly up to World War II, witnessed increased regulation and control of the main domestic transport industries (Button & Gillingwater, 1986). Arguments invoking public interest considerations dominated debate. While the pattern was general to most modes and most countries, as far as aviation was concerned regulatory legislation such as the Civil Aeronautics Act (1938) in the USA, the Air Navigation Act (1920) in Australia, the Trans Canadian Airlines Act (1937) in Canada and the Civil Aviation Act (1946) in the UK demonstrate the universality of controls in this industry. (They applied even in countries which are, in terms of the dichotomy set out later, in Section 4.5, normally viewed as favouring commercial approaches to transport policy.)

Regulation of international aviation had its birth at the Paris Convention of 1919 which accepted that states have sovereign rights over the air space above their territory. This immediately involved national governments in the regulation of the industry. They now enjoyed an agreed right to control what was rapidly becoming a valuable natural resource. Initially, international aviation was largely left under the control of a set of *ad hoc* arrangements between countries. Given the technical factors limiting flight duration and the commercial viability of services providing long range air transport at the time, it seems unlikely that this seriously impeded progress in the industry.

The period after the end of World War II, and in particular following the failure of the Chicago Conference of 1944 to reach a comprehensive, multilateral agreement on 'freedoms' of the air mainly because of fears that the US would dominate a completely free market, saw the separate states of Europe continue to exercise total sovereignty over the control of their airspace. The problem was that the Chicago convention reached agreement on some of the freedoms of the skies – namely the first (the right to fly over another country's territory) and the second (the right of an airline to make a technical landing to, for example, refuel, but not to pick-up or set-down passengers) – but not on the remainder. The third freedom relates to the right of an airline from country A putting-down passengers, picked-up in the country of registration, in country B. The fourth freedom relates to the right of an airline from country A to pick-up passengers in country B for off-loading in country A. The fifth freedom relates to the rights of an airline not registered in either country A or country B to carry passengers between them. Since no multilateral agreement was reached on these latter freedoms a series of bilateral agreements emerged with IATA acting as a forum for discussion and co-ordination.

As a result of the inability to reach agreements on a multilateral basis, a system of very rigid sub-markets emerged in air transport with the major routes generally being shared between the national carriers of the countries concerned (see Doganis, 1985; Kasper, 1988). Fares normally had to be agreed by both countries involved and, as can be seen from Table 4.5, multiple designations of routes (with more than one carrier from a country) were very rare as was the granting of fifth freedom rights. In most instances the capacity offered by each country was limited to 50% with revenue-share pools a common practice. The general view of the situation prevailing at the beginning of the 1980s was well summarized by the House of Lords Select Committee on the European

Table 4.5 European Communities air services (return journeys) 1987.

From	Routes	With multiple designations	With fifth freedom rights
France	153	2	11
Greece	39	0	6
Italy	99	1	9
Spain	108	2	5
Belgium	45	1	4
Denmark	37	0	7
Ireland	43	2	3
Luxembourg	15	0	2
Netherlands	65	7	9
Portugal	33	2	1
United Kingdom	178	24	14
West Germany	173	7	17

Source: Commission of the European Communities data.

Communities (1980), *viz* '... the main consequence of this system for allocating routes, fixing fares, pooling revenue and so on has been until recently the virtual elimination of competition in fares on scheduled services'. Equally, many countries, as we noted above, exercised strong controls over their own domestic aviation and, in several cases, the preferred national carrier enjoyed monopoly rights. Certainly cabotage was not permitted.

Although the predominant system of bilateral agreements has always placed constraints on the liberalization of European aviation, the actual importance of these constraints in the past may sometimes have been overstated (see also Button & Swann, 1989a). Aviation is, by its nature, a very flexible industry which has tended to attract innovative management. Ways of circumventing the worst of the constraints were, therefore, often found. At the same time, market entry and fare controls were, in several different and indirect ways, gradually relaxed over the years and 'alternative' aviation markets were developed outside of the regulatory system which applied to scheduled activities.

Both domestic and international forces have played some part in these changes. While the forces at work have interacted, and the policy changes and developments which have taken place should not strictly be treated in isolation from each other, we can examine some of the broad trends under a number of headings.

The liberalisation of bilateral agreements

A trend towards greater flexibility in bilateral arrangements, and here we temporarily extend ourselves beyond internal European traffic and embrace international aviation betwen European states and outside countries, can be detected in the late 1970s. In particular, the US actively began to seek ways of liberalizing its bilateral arrangements with European nations (Kasper, 1988;

Gomez-Ibanez & Morgan, 1984). This process, which was linked with the domestic deregulation taking place within the US, was first initiated in the USA during the mid-1970s. In particular, efforts were made after 1977 to extend the new liberal philosophy to the rather heavily regulated international operations on trans-Atlantic routes. The adoption of a more liberal posture on the international front, designed specifically to 'encourage vigorous competition' – the 'Open Skies' policy – was, however, only achieved after a rather fraught period of negotiations with the UK which had initially complained about unfair, restrictive US practices. The speed of liberalization was increased through the impact of a 'beachhead' strategy adopted by the US which was designed to penetrate one national market at a time and then force liberal agreements on others by virtue of the threat of traffic diversion. Revised treaties were rapidly signed, for example, with Belgium, the UK and the Netherlands. All represented, to varying degrees, a more liberal approach to market entry.

These changes, by virtue of both the inter-related nature of trans-Atlantic sub-markets and the possibility of using surfaced based transport within Europe to widen the range of airline options available to passengers, then became one stimulus for liberalization of bilateral air service agreements between European states. In one way, however, their effect was rather more psychological than actual. While scheduled services across the Atlantic had formerly been heavily regulated there was a thriving charter market. In many ways liberalization, therefore, while certainly stimulating more traffic, may be seen as causing a switch out of charters and onto scheduled flights. In 1976, for example, scheduled flights accounted for about 75% of passengers, by 1981 this figure had risen to 94% and by 1986 to 95%.

Partly as a result of changes on the trans-Atlantic sector but also for internal reasons, developments also took place with regard to some intra-European bilaterals. Thus while it took time for the 12 members of the EC, for example, to make progress towards liberalization on a Community-wide basis, liberalization was possible between pairs of countries. The liberalization of a number of agreements within Europe has been important both in its own right as a means of providing more efficient services and in its role of providing a demonstration that liberalization does not necessarily mean chaos in the market place.

Since the mid-1980s a number of steps have been made in this direction. The UK–Netherlands bilateral of 1984, for instance, opened access to routes by allowing any airline from either county to offer services between them. This measure was complemented by the removal of compulsory consultation on fares which where now to be determined by the country of origin. Further fare liberalization followed in 1985 when airlines were left free to set their own fares unless both governments expressed disapproval. Bilateral liberalization has followed, albeit to varying degrees (see Table 4.6), on a number of other routes e.g. UK–West Germany; UK–Belgium; UK–Luxembourg, UK–Switzerland, UK–France, UK–Ireland, UK–Italy, and UK–Spain – see Pelkmans (1986) and Barrett (1987). A more, albeit rather limited, agreement has also been reached between France and West Germany which has liberalized the capacity

Table 4.6 Liberalized UK bilaterals with other European countries.

Country	Route access	Liberalization of: Capacity constraint	Tariff constraint
Netherlands (1984)	Yes	Yes	No
Netherlands (1985)	Yes	Yes	Yes
West Germany	Yes	Yes	Limited
Luxembourg	Yes	Yes	Yes
Belgium	Yes	Yes	Yes
Switzerland	Yes	Yes	Limited
France	Limited	Limited	No
Spain	Limited	Limited	No
Italy	Limited	Limited	No

constraint and airlines from both countries can, with some caveats, serve any point in the other country.

The liberalized bilaterals have both resulted in greater flexibility in the types of services offered on the routes involved and served to demonstrate that excessive instability need not arise under freer market conditions. Initially, however, the impact was limited (Commission of the European Communities, 1987a). Traffic on UK/Netherlands services, for example, actually grew more slowly after the change than on many other European routes (10.8% in 1986 compared with a range of 6.4% to 17.5% on other UK–European routes). This may, however, be explained in terms of the relatively liberal regime that operated prior to 1984 and the inherently small size of the UK–Netherlands market. The liberalization of the UK-Ireland bilateral in 1986, which allowed some capacity increase and removed fare constraints, has produced more dramatic result. While fares had risen by 72.6% between 1980 and 1985 (compared, for example, with retail price increases in Ireland of 41.5%), from 1985 to 1989 the unrestricted return air fare fell from IR£208 to as little as IR£70. By 1989 the number of daily flights between London (including Luton) and Dublin had risen to 82 compared with 32 in 1986, and passenger numbers had risen from 860,000 per annum to 2.5 million. The total traffic between Britain and Ireland rose from 1.85 million passengers in 1985 to about 4.2 million in 1989 after a decline of some 100,000 passengers had been experienced between 1978 and 1985.

Non-scheduled operations

While scheduled services have been subjected to varying degrees of regulation, there has existed much more flexibility with regard to charter operations and inclusive tours. While IATA had in its early years attempted to control charter fares, its influence in this area dwindled with time (International Air Transport Association, 1974). As the European Civil Aviation Commission (1981) has put it, '... most member states, irrespective of their approach towards the regulation of scheduled services, have a comparatively liberal approach towards the regulation of non-scheduled operations'.

Non-scheduled services were defined as long ago as 1952 by ICAO as services which do not, '... operate according to a published schedule or on a regular basis so that a schedule is clearly identifiable'. Subsequent, national definitions, have sometimes tended to be somewhat tighter than this especially if governments wish to protect their scheduled operators. Although the exact details of definition and regulations vary between countries, the rules governing charter services normally include: prohibitions on airlines selling tickets directly to the public; tickets can only form part of a larger package which normally includes accommodation; the length of maximum/minimum stay must be stipulated; freight cannot be carried on the same aircraft as passengers; and the service must not adversely affect scheduled services (National Consumer Council, 1986). These are obvious constraints on operations, but the important point is that once operators comply with them, market entry is unrestricted and fares unregulated.

While the background to the actual system of charter regulation in Europe can be traced back to the 1956 ECAC, Multilateral Agreement on the Commercial Rights of Non-scheduled Air Services in Europe, the 1980s saw an exploitation by carriers of the greater flexibility afforded through offering non-scheduled operations rather than scheduled services. For example, by 1985 over 75% of passenger travelling, UK–Spain, UK–Portugal and UK–Greece went on non-scheduled services.

Domestic aviation policy

Finally, the 1980s saw limited reforms of domestic aviation policy in some countries. The UK is perhaps the most advanced in this sense. The system of tight regulation of entry and price controls which were administered by the Civil Aviation Authority (CAA) since 1946 have gradually been relaxed. In particular from 1980 the Civil Aviation Act required the CAA in its licensing activities, 'to have regard in particular to any benefits which may arise from enabling two or more airlines to provide the service in question'. Further *de facto* liberalization followed. In 1982, for example, the CAA began granting licences to allow larger airlines to compete with British Airways on domestic routes out of Heathrow. This was followed by the CAA licensing statement of 1984 (Civil Aviation Authority, 1984b and the Annex by Barnes in Organization for Economic Co-operation and Development, 1988) after which the importance of potential competition as well as actual competition was explicitly recognized as in the interests of users (Civil Aviation Authority, 1987, 1988). As a result, area-wide licensing facilities were initiated permitting airlines to fly between any two points in the UK. Domestic fare controls have also been relaxed and the CAA now acts more as a monitoring agency and only intervenes when it suspects predatory pricing or the exercise of monopoly power in a particular market segment.

While such reforms are important, it should perhaps be made clear that there is little evidence of UK types of policies being widely adopted for domestic aviation in other European countries. Indeed the recent refusal of the French authorities to allow its long-established intercontinental carrier. UTA, access to European routes, and retaining them as a monopoly preserve of the state-owned Air France, tends to be a more normal pattern of behaviour.

In addition to reforms in regulatory policy there has also been a limited degree of state withdrawal from airline companies themselves. The most important of these was the privatization of British Airways in 1987 – see Ashworth and Forsyth (1984) for a discussion of how this fitted in with UK aviation policy more generally. One should also note that the British Airports Authority (BAA) was also privatized in the same year. While this clearly removes aviation from direct ministerial/political control, and therefore may well offer scope for greater economic efficiency, it has meant that a certain amount of new regulation has been introduced for public interest reasons. In particular, BAA owns the two main London international terminals and, hence, is in a position to exercise monopoly power. Of course, a different divestment strategy designed to produce competition between the airports may have reduced the need for strong regulation.

Changes have, therefore, taken place over the past decade in the aviation sector. These, as we have noted in the above account, have generally been evolutionary rather than revolutionary.

4.4 Pressures for further reform

As we have seen, the regulatory system in Europe has not remained static. Gradual change can be detected in certain spheres and airline operators have been inventive in many cases in circumventing some of the more restrictive regulations; some of the 'accommodation' offered as part of charter packages, for example, can at best be described as minimal. There has, nevertheless, been a considerable build-up over the past decade of pressure for greater and more rapid liberalization of aviation markets in Europe. Some of this pressure has come from within Europe itself but there have also been wider influences at work. It is perhaps useful to examine the major forces for more dramatic change under several broad headings.

The movement towards a single market

The early 1980s saw the EC reach something of a watershed in its development. It expanded as Greece and then Spain and Portugal joined adding new dimensions to the problems to the development of EC transport policy. More generally, however, it had reached a stage where discussions concerning its long-term direction were required. It had essentially become paralysed by internal disputes and new initiatives were needed (Swann, 1988b). It was as a result of this that the Single European Act was agreed in early 1986. This did not contain provision for a European Union as some countries hoped but rather set down the basis for establishing a Single European Market by 1992 (Commission of the European Communities, 1985). This had implications not simply for trade and commercial policy but also for transport policy within the 12.

The Common Transport Policy of the EC had not developed at the pace initially hoped for (Erdmenger, 1983) and aviation policy had always occupied a rather awkward position within the overall framework. The move towards a Single European Market put pressures on the EC not only to resolve such traditional problems as the granting of greater freedom for road haulage operators to offer genuine EC-wide services and the simplification of border crossing formalities, but also to integrate aviation within the overall framework of the EC's policies.

Linked to this in some ways was the recognition that international transport was the fastest growing transport sector within Europe and that the existing *ad hoc* regimes of regulations were incapable of handling the demands which were being placed on the industries involved. In terms of aviation, capacity was proving inadequate to meet existing demand and would certainly fall short of that required to meet the demand forecast to the end of the century. Within the EC such pressures manifested themselves in the form of actions by operators and individuals to bring about reform. The specific implications of these issues are dealt with in more detail in section 4.6 which reviews EC aviation policy in the 1980s. The point of emphasis here is the wider pressures of bringing about the Single Market which were focused on aviation.

Developments in the aviation market

The aviation market in Europe is likely to be the subject of continuing change on both the demand and supply side.

With regard to demand, forecasts vary but many official projections within Europe (e.g. that of Eurocontrol and ECAC) suggest that air traffic could double by the end of the century. The Department of Transport forecast for UK–Continental traffic (which accounts for about 50% of European traffic) is an increase of something above 6% per annum giving a slightly lower overall prediction than most other bodies. Much of the growth is a product of growing affluence and the availability of more leisure time to take holidays. In addition, however, the creation of a Single Market will itself generate more international business travel.

On the supply side, while much of the recent growth has been due to more people travelling there have also been important changes in the type of aircraft being used. It had been anticipated that congestion would be contained by the use of wide-bodied jets, such as the Airbus 300 and 310 and the Boeing 767, to carry the additional numbers of passengers. In the event, the number of aircraft movements has risen faster than the number of passengers as airlines have opted to deploy smaller aircraft (e.g. the British Aerospace BAe146). This trend has major implications for both air traffic control and airport provision and, in a much wider context, for the noise nuisance and atmospheric pollution associated with air transport.

The growth to date has been mainly in the context of a relatively regulated market and with adequate infrastructure. This means that, unless the same regulations ensure a good match between supply and demand in the future, both for the market as a whole and for sub-markets within it, then frustrated demand will inevitably develop with all its associated problems. The frequently-cited forecasts by ECAC and others, for instance, that there will be a doubling of annual aircraft movements between 1987 and 2000 is conditional on airport and air traffic control system capacities not acting as constraints. Here there is a contradiction. Liberalization of European aviation will, it is argued, permit users to select flights more rationally and enhance consumer welfare but equally it will put intolerable strains on the infrastructure unless it is considerably expanded (Organization for Economic Co-operation and Development, 1988). In the short term there is also the problem of rationing the infrastructure which is available in the absence of economic prices being charged for its use. Hence while there is market pressure for speedy reforms to aviation regulations, policy makers favour a more gradual approach while facilities are developed and modernized and allocation strategies decided upon. We return to these problems later.

Set against these pressures for increased efficiency from the aviation system is the possibility that some factors outside the aviation market may act to contain some of the growth in demand. Most forecasts of air traffic, for example, assume that current short-term trends in factor prices and income growth will continue. Recent history has been characterized by relatively fast economic growth and low fuel prices (see Table 4.7). There is some doubt as to whether both will persist to the end of the decade, indeed there is some evidence that economic growth is already slowing in many EC states. If this does happen then pressures for rapid liberalization will recede although this is only likely to be a short-term respite.

Table 4.7 European spot jet fuel prices.

Year	Price (US cents per US gallon)
1982	96.5
1983	83.9
1984	79.0
1985	79.7
1986	48.5
1987	52.4
1988	47.3
1989 (April)	55.5

Demonstration effects

Demonstration effects have also played their part in forcing change. The past decade has witnessed considerable liberalization of transport and other markets around the world (Button & Swann, 1989b). Perhaps, rather chauvinisticly, one could suggest that the stimulus for much of the reform in transport policy lay in the success of the UK's 1968 Transport Act, and its freeing of the road haulage industry from economic regulation, but more realistically the essential demonstration effect came from the USA and, in particular, from the outcome of the 1978 Airline Deregulation Act (Bailey, 1985; Levine, 1987; Kahn, 1988; Keeler, 1990 and Pickrell, chapter 2 in this volume).

The experience of the US in liberalizing its domestic aviation industry demonstrated that the preceding period of extensive market regulation had stifled the natural development of the market; led to excessive fares; fostered inefficiency and limited consumer choice. The US inter-state aviation industry had been regulated since 1938 with fares controlled and routes licensed on an individual basis. By the mid-1970s it had become clear that such a regime was not maximizing economic efficiency. In particular, it impeded the natural growth of 'hub-and-spoke' operations and consequently meant that economies of density and scope could not be fully exploited. In contrast, costs can be considerably reduced by employing key hubs as consolidation points for flights with routes radiating out from them (Morrison, 1989). The replacement of linear, direct services by hub-and-spoke operations, because of the indirect routings and probable changes of plane required at a hubs, increases journey times for passengers but it also reduces airlines' costs (and in a competitive environment, also fares) and increases the range of flight possibilities available. The results include: average load factors per plane up from less than 55% in 1978 to about 62% in 1988 and passenger boardings up from 275 million to 455 million while air fares, adjusted for inflation, fell by 21% over the period (Labich, 1989).

Inevitably the US experience and its fruits of lower fares attracted the attention of Europeans. Much of the interest lay in the dramatic reduction which occurred in US domestic scheduled fares and in comparisons of these with the much higher levels prevailing in Europe – for comparisons see, Civil Aviation Authority (1983) and Barrett (1987). The House of Lords Select Committee on the European Communities (1980), for instance, pointed to instances of fares in Europe being double those for comparable services in the USA although these were extreme examples and, in particular differences in return fares were in general somewhat smaller. The Committee explained the difference thus, 'The interests of consumers appear to be sacrificed to the prestige of flag carrying national airlines and the protected environment in which they operate'.

Of course the US reforms have not been trouble-free and the impression gained in Europe may be seen as one viewed through rose-tinted spectacles. Airlines in particular have responded to the new situation by trying to reduce the edge of competition in the market (Levine, 1987). Because of a concurrent relaxation of anti-trust policy, for example, mergers have occurred between airlines limiting competition between them on particular routes (although because of increased hubbing not necessarily between various origin and destination pairs) and over landing slots at some key, hub airports. In the latter context, for example, by purchasing Ozark Air Lines. TWA gained control of 83% of the traffic at St Louis, Lambert Airport while the merger of Northwest and Republic meant the resultant company provided 77% of the flights into Minneapolis and 62% of those at Detroit (Labich, 1989).

Mergers are not the only device used to blunt the edge of competition. The details set out on computer reservation systems (CRSs) used by travel agents, but owned by airlines, for giving customers flight information and for booking their seats, have been manipulated to favour the parent company. Even though one of the final acts of the Civil Aeronautics Board before its demise in 1985 was to initiate controls over such practices, some 60% of travel agents still write tickets on systems provided by United or American. Bonus schemes and other incentives suggest that the 'halo effects' that result, bias the system even if the information held within the CRS programme is objectively presented. Frequent-flyer programmes, giving bonus flights to loyal customers, have been deployed as a defensive weapon to encourage existing passengers to stay with the airline currently favoured and as an offensive device to attract flyers away from companies not offering such a perk.

Despite these difficulties, the general view seems to be that liberalization in the US, has, in overall cost-benefit terms, yielded a positive rate of return. While market imperfections exist and the post-reform market hardly corresponds to the ideals established as bench marks for efficiency by economists, the distortions associated with the market are seen as less damaging than the government failures which accompanied the previous, heavily regulated regime (Keeler, 1990). Policy makers in Europe, therefore, have seen this as providing a case for liberalization of markets there, albeit in a modified form designed to contain the major problems which have arisen in the USA.

Developments in economic theory

Accompanying the changes in the regulation of domestic aviation within the USA and, indeed liberalization of other transport markets within Europe, has been the emergence of powerful new economic theories regarding the ways markets work. While it is difficult to explain some of the past changes of policy in terms of the widespread acceptance of these new ideas, for example reforms of UK road haulage regulations in the 1968 Transport Act were well in advance of the new ideas and those responsible for the deregulation of US aviation seem not to fully subscribe to all of them (e.g. Kahn, 1988 and Levine, 1987) – they are nevertheless potent forces in the more recent debates about European aviation. Two particular developments in economic theory are of special importance.

The notion of regulatory capture

By the mid-1970s there was a growing body of opinion, led by the Chicago School of economists, that economic regulation might not always be initiated with the public interest in mind. Further, even if originally it was, it might subsequently cease to serve that public interest. The reason for this being that the legislators and/or the regulators might be *captured* by those whom it was intended should be controlled and the regulators might well have a vested interest in expanding the system to meet their own aspirations for power (Stigler, 1971).

A major difficulty with any form of regulation is to ensure the regime is appropriately updated as conditions change. Moreover, the problem is that those with the specialist knowledge needed to carry this through are normally those actually involved in the industry – hence they have the power of manipulation. In these circumstances, the Chicago School argues, the onus of proof should very much be on the shoulders of those advocating regulation and where there is doubt a liberal policy is to be preferred. There was considerable debate in the US about the applicability of such arguments with regard to the transport, and particularly aviation, industries (Keeler, 1984) and this permeated thought within Europe.

Contestability

The second important development in economic thinking concerns the power of potential competition. Traditionally market intervention in aviation has been justified as being important to contain the power of monopoly suppliers of services, especially on thin routes where efficiency may well mean only one airline could operate viably. *Contestability* theory argues that provided there is free entry to a market and exit from it (i.e there are no 'sunk costs'), then the threat of potential competition will deter any efforts by a monopoly or quasi-monopoly operator to exploit air travellers (Baumol *et al.*, 1982). If there are no sunk costs associated with providing a particular service then the simple threat of hit-and-run entry would deter incumbent operators from charging sub-optimally high fares.

While perfectly contestable markets are unlikely to be achieved, any more than perfectly regulated markets ever exist, a high level of contestability is possible in the aviation industry (Bailey & Panzar, 1981). Aircraft, for example, are easily switched between routes at a negligible cost. Advocates of liberalization argue, therefore, that removal of entry, exit and pricing regulations would, in these circumstances, be preferable to existing regulatory regimes. In these terms Keeler (1990) has implied that aviation markets can be made 'workably contestable' in the sense that liberalization generates a more efficient, if not perfect, system than one of economic regulation.

4.5 Constraints on the drive for liberalization

While advocates may advance strong arguments for the liberalizing of aviation markets within Europe, there are practical problems to be considered. Some of these have been hinted at already. Further, not everyone accepts the arguments for reform, or at least not on the scale of the most liberal-minded advocates, and point to weaknesses in the cases being advanced.

In this section we look at a number of particular issues which may slow any progress towards greater liberalization of aviation in Europe. First there is the question of the ability of European nations to act in concert in arriving at a liberalized market in aviation. While there are clear problems involved of the attitudes of Easter bloc states *vis-à-vis* Western Europe there are also serious differences in attitude and philosophy between even members of the EC.

The second issue is whether we can really take the US experience as a demonstration of what will happen in Europe over say the next decade. This involves considering whether liberalization could or should go as far as it has in the US. If we cannot adopt the US example, then it becomes necessary to have either an alternative framework for reference – or at least use the US one with caution – or to adopt a rather doctrinaire stance and possibly pursue liberalization as being beneficial in itself.

The third issue we review is a much more practical one. All forms of modern transport require extensive and appropriate infrastrcuture. Aviation is not simply about airlines it also involves airports, air traffic control, air space, etc and actions in these fields must inevitably influence the nature of the overall aviation market. We consider, therefore, important infrastructure constraints which may impede the liberalization processes or at least influence the detailed form which they take.

Divergent philosophies

Within the EC there are overlapping philosophies and approaches to economic regulation which extend into the supply of aviation services. The existing patchwork pattern of controls, over such things as market entry, fares and conditions of operation, have obviously grown with time. They are to a large extent a reflection of these differences. Countries such as France, Spain and Greece, where domestic aviation is relatively important, have a tradition of heavily regulating both entry and fares and this tradition has extended to their views of international aviation policy. This degree of regulation has frequently been justified by governments in terms of serving the public interest by, for example, ensuring market stability, maintaining safety standards, protecting the public from monopoly exploitation and providing a comprehensive network of services (Organization for Economic Co-operation and Development, 1988). Air transport, from this perspective, is being viewed as a public service where regulation is seen as necessary if reliable, regular services are to be offered at the lowest possible costs consistent with a reasonable rate of return being earned by carriers.

These controls also serve, perhaps less explicitly, as important instruments for the protection of other aspects of the national interests by maintaining flag carriers which meet wider economic and military criteria. Exporting aviation services can, for example, represent an important element of 'invisible' earnings from foreign trade. In addition, there is the question of status and market presence. In many countries (e.g. Greece) aviation is, for instance, provided through statute by national, state-owned airlines. Such direct controls act not only to influence the air transport services being provided but can also be deployed to regulate the purchase of aviation equipment which can also form a major item in foreign trade accounts.

Such a regime of *ad hoc*, state-based regulations and controls is unlikely to generate the most efficient aviation system for Europe. While some countries may benefit because of their bargaining position or through historical accident, overall it will tend to protect inefficient operators and distort the overall pattern of services being provided. They effectively act as a restraint on trade and have associated with them the same undesirable economic implications as other restraints such as tariffs. The problem is that countries which have well entrenched systems of market controls, even if appreciative of the probable adverse implications of this for the overall welfare of the Community, have effectively cushioned their airlines from competition and, in consequence, they would find it hard to compete in a more market orientated environment.

The relevance of the US experience

There are strong pressures for liberalizing European aviation, and supportive evidence both from economic theory and the experience of the USA indicates that, on balance, a liberalized market may prove more efficient. In practice, however, change is not easy. In particular the European aviation market differs in several major respects from the US domestic market – see also Civil Aviation Authority (1984a); Pryke (1988; 1990); Pelkmans (1990); Tucci (1985); Association of European Airlines (1984) and Button (1989a). These differences tend to indicate that liberalization in Europe may not produce an identical result to that experienced across the Atlantic. For example one can point to the facts that:

Domestic/international traffic split
The deregulated US market is domestic whereas that within Europe is predominantly international – some 80% of European flights are cross-border. This means that the majority of European carriers are essentially international airlines. The creation of a Single Market within the EC by 1992 (Commission of the European Communities, 1985) will still leave many European nations outside the liberalization process and even within the EC not all constraints are

likely to be removed given the variety of economic, social and political objectives which underline nations' aviation policy. The changes in the US were sudden and uniform, those taking place in Europe are gradual and, to some extent, variable.

The non-scheduled market

The internal structure of the European market differs significantly from that of the USA. Europe has a substantial charter market which did not exist on the same scale in the US prior to deregulation. Further, not only have non-scheduled operators the scope for relatively easy market entry, albeit with conditions attached, but many of them have relatively new fleets and have an established market image. Unlike the situation prevailing in the US until the mid-1970s, this means there are powerful countervailing forces in many markets already restraining the actions of scheduled operators. Indeed, a significant amount of scheduled capacity is already filled by passengers travelling on discount tickets.

Market size

The size of the European market is significantly smaller than the domestic US market (Pelkmans, 1986). The average route length in Europe is some 750 km whereas in the US it is 1300 km. Taking the top 75 routes in Europe only 17 have a flight time of two hours or more and for 10 of these flight time is between two and two and a half hours (Pryke, 1988). US flights tend to be much longer *and this is important*. If flights are short then there is much less scope for hubbing because any time spent changing plane during a trip takes up a *relatively* long time. In terms of market competition, it means that indirect flights in Europe, even if fares are lower, are seldom going to offer effective competition to direct services as has occurred in the US.

The nature of the European market, and especially the small number of routes with heavy traffic volumes, also means that the eventual degree of competition which is likely to emerge post-liberalization is almost certainly less than that found in the USA. Pryke (1990), for instance, deploying a set of assumptions regarding levels of hubbing and demand, suggests that while liberalization in Europe will reduce the degree of monopolistic power over many routes (e.g. on short-haul routes, while single-carrier supply will remain on about 48% of services, the number of two carrier routes will fall from 32% to less than 25% as multiple supply expands), the eventual outcome will fall short of the US situation (where only 38% of routes have one supplier and 25% have two suppliers).

Production costs

Production costs are different in Europe to those which prevail in the US. While it is difficult to make direct comparisons, there is some general evidence (see Table 4.8) that scheduled airline costs are higher and productivity lower in Europe than in the USA (see also, International Air Transport Association,

Table 4.8   Airline labour costs and productivity (average per employee, 1987).

Carriers	Pilots/ co-pilots	Labour costs ($thousand)		Productivity*
		Other cockpit staff	Cabin crew	
Eight US majors	92	40	28	1.6
BA/BCal (Britain)	65	48	19	1.1
Lufthansa (West Germany)	na	130	40	0.8
SAS (Scandinavia)	na	103	41	0.6
UTA (France)	164	119	45	0.8
Alitalia (Italy)	na	93	59	0.7
Iberia (Spain)	109	80	37	0.7

* Million revenue passenger kilometres.
Source: McGowan, F. and Seabright, P. (1989), Deregulating European Airlines, *Economic Policy*, **9**, 283–344.

1984; Forsyth *et al.*, 1986; Sawers, 1987; and Barrett, 1987 for earlier data). European airlines are confronted with some costs which are outside their control and are higher than those encountered by their US counterparts. For example, IATA has estimated that landing fees in the US represent between 10% and 30% of the European level. Fuel is another example of an exogenous factor price difference. But even allowing for this, and the differing nature of the European market discussed above, some of the higher costs stem from lower productivity rather than generally higher unit input prices. While deregulation in the US resulted in the substantial reductions in labour costs (Pryke, 1987) and enhanced productivity, mainly brought about by wage reductions, labour shedding, changes in working practices, etc, it seems unlikely that the somewhat different attitude towards labour relations prevailing in most European countries would permit the same thing to happen there – or, at least, not so rapidly and dramatically.

Computer Reservation Systems
The CRS systems in the USA are owned by the largest airlines (e.g. United with Apollo and American with Sabre) while in Europe they are owned by a number of airlines (e.g. Galileo is owned by British Airways, KLM and seven other airlines). They are, therefore, unlikely to be open to quite the same degree of potential exploitation as the US system (see, House of Commons Transport Committee, 1988). In particular, US owners are able to access rivals' databases while the European systems are designed to prevent such 'keyhole' activities.

Ownership of airlines
While the US aviation industry is entirely in private hands, there is substantial, although very slowly contracting, public sector participation in the ownership of European airlines (see again Table 4.2). In many countries there is a 'preferred vehicle', namely a dominant airline which is normally either partly or entirely publicly owned and is seen as the instrument for advancing national aviation policy. While several US carriers have gone bankrupt, public sector

involvement inevitably affects the way an airline is treated and it is difficult to conceive of a government-owned carrier being allowed to go bankrupt in a competitive European aviation market.

Inter-modal competition

There is substantial inter-modal competition in Europe over medium distances, especially from high-speed train services such as the French TGV system (Seidenfuss, 1983). The US rail system, save for some services on the North-East corridor, is essentially a freight system – less than 0.05% of US passenger traffic is by rail. The situation is different in Europe. Some 13% of passenger-kilometres done in the EC are by rail. The importance of rail is particularly pronounced for journeys of between 400 and 1,500 km. This is in part because the rail lobby is a very powerful one in Europe and hence railways tend to be heavily subsidized (Seidenfuss, 1983). In themselves, though, subsidized fares are not enough to explain rail patronage. A second point is that the high quality of the network and local transport provision makes rail a more attractive travel mode than in the USA. Indeed, as the data provided above suggests, rail services can often offer a viable practical alternative to air transport, on the basis of door-to-door journey times, for trips up to 1,500 km.

The expansion of the high-speed rail network in Europe, the completion of the Channel Tunnel, and so on (Button, 1989b) suggest that this competition will expand to more routes over the next decade. There is already evidence that the development of such services can affect air travel demand. The ability of high-speed trains to attract air passengers has been proven first in Japan at 240 km/hour then in France at 270 km/hour. In the latter case, for instance, Figure 4.1 shows the growth in demand for air travel on the main French domestic routes. The opening of the TGV rail service between Paris and Lyon has clearly contained air traffic growth on this route (air has lost about 50% of the traffic it might have expected without TGV competition) as has the Paris–Geneva service.

High quality roads in Europe, especially coupled with faster permitted driving speeds in many countries, also means that road passenger transport competes rather more effectively on some corridors (especially for journeys of less than 1,000 km) than would be the case on comparable routes in the US. The gradual evolution of an EC infrastructure fund coupled with national investment programmes (albeit differing in scale between EC members) also means that the European road network will continue to develop in the future.

The advantage of hindsight

European policy makers already have the US experience to guide them and they are, therefore, likely to react against some of the perceived difficulties which have been encountered in the US. The Directorate General for Competition of the EC, for example, would seem to be fully aware of the potential problems airline mergers may cause under a more liberal regime and the latent market power which exists through flight code sharing and domination of CRS systems

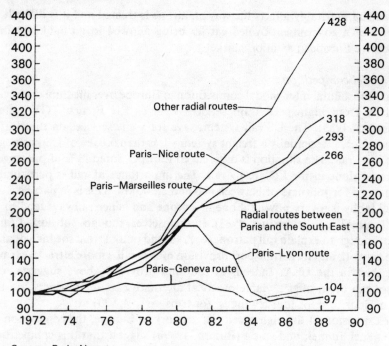

Source: Paris Airport.

Figure 4.1   Air travel on the main French routes.

(Argyris, 1989). The implementation of policies to counter potential problems, however, will itself cause reaction among operators as they naturally seek some degree of shelter within the more competitive environment. Given the different institutional constraints confronting them, the airlines may well behave differently to their US counterparts.

While there are these differences between the US and European situation, from a policy perspective it is unlikely that the adoption of US style liberalization would result in a situation emerging in Europe which would be markedly different to that on the other side of the Atlantic. The key point is that there are institutional and social reasons which are likely to preclude the full adoption of US style 'deregulation'. It is also unlikely, given the institutions involved, to come about very rapidly. The problem, therefore is to judge which lessons are particularly applicable to the context of Europen reform and to learn from those.

Infrastructure constraints

Besides these quite major differences between the US and Europe there is another major technical problem to overcome – the adequate provision and allocation of aviation infrastructure in a liberalized market.

While demand growth, and the need to cope with it efficiently, is seen by many as a justification for market liberalization, there are constraints on the supply side that may become more serious if traflc growth continues unabated. There is a perceived need, therefore, to make the most efficient use of the infrastructure available and ensure that investment in additional facilities is made to meet future demands. The latter is seen as a particular problem since it takes about 10 to 12 years to design and build a large airport. Indeed, one argument for gradualism in any process of deregulation is the need to ensure that the basic infrastructure is adequate to cope with the additional traffic which would almost inevitably accompany a more liberal regime.

One of the constraints which has been used in the past as a justification for restricting the unfettered growth of European aviation has been the inadequacy of supply of infrastructure available. The capacity of the existing air transport infrastructure in Europe, and the way in which it is operated, clearly places a constraint on the longer term development of aviation (Commission of the European Communities, 1989). Indeed, many would point to the fact that in the first half of 1989 some 21.3% of flights in Western Europe were delayed for longer than 15 minutes compared to only 13% in 1986 and say the problem is already with us. Increased market liberalization is, in particular, likely to have implications for both airports and air traffic control systems. Some plans are, however, already afoot to meet the changing situation.

Airport capacity

With respect to airports, the US experience suggests liberalization is likely to have two significant types of effect. First, it will lead to more air travel as competition stimulates efficiency and reduces fares. As we have seen, the impact in Europe may, however, be less than some people anticipate, in part because gradual liberalization has already been introduced on a bilateral basis on many routes but also because, unlike the pre-deregulation situation in the USA, a considerable charter market already exists and liberalization will simple see some transfer from this to the scheduled market. Second, as witnessed in the US, efficiency in aviation, comes from the maximum use of hub-and-spoke styles of operation. There is also some indication that the EC accepts this latter point as was illustrated in a speech by the Commissioner for Competition in July 1988 when he envisaged that the long list of airports excluded from the 'hub to regional' provisions will be substantially reduced.

A substantial amount of European air traffic already hubs (see Figure 4.2) – in part because of the relatively large amount of inter-continental traffic involved – but there are mounting capacity constraints at many of the major hubs. Of the 46 largest airports in Western Europe, 10 are operating at or around capacity and it is predicted that a further 13 will join this list by 1995 with further problems emerging at the end of the century (see Figure 4.3).

The rather more gradualist approach adopted towards liberalization in Europe than in the USA may be seen as a conscious effort to contain the growth

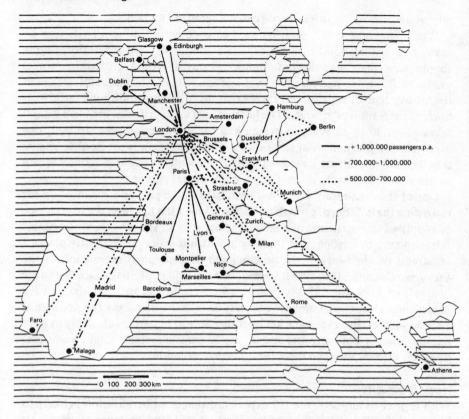

Source: Community of European Railways (1988), *Proposals for a High-speed Network* (Paris, CER).

Figure 4.2   Main air traffic flows in Europe.

at these hubs and make more intensive use of spare capacity at provincial airports. There is an appreciation in several countries that more capacity will be required in the near future and major expansion plans have been drawn up for a number of European airports. In the UK, for example, a new terminal is under construction at London-Stanstead, Italy plans to expand Rome-Fiumicino, the Netherlands is expanding terminal facilities at Amsterdam-Schiphol, Czechoslovakia plans a new terminal at Prague-Ruyne and Belgium has plans to substantially expand Brussels Airport. Munich II, however, is the only new major airport which is under construction although a greenfield site for a future Oslo regional main airport to become operational at about the turn of the century has been selected at Hurum.

If further new capacity is to be provided then this will require funding and in the meantime existing capacity if it is to be used optimally will require that traffic

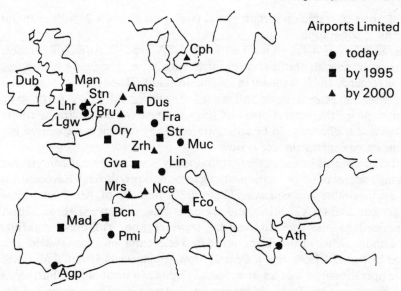

Source: Association of European Airlines (1987), *Capacity of Aviation Systems in Europe. Scenario on Airport Congestion* (Brussels, AEA).

Figure 4.3 Congested airports in Europe.

restraint measures are introduced. The cost of additional infrastructure, especially if the full social and environmental effects are taken into account, is likely to at least partly offset potential efficiency gains from market liberalization. Equally, rationing of existing facilities by relying on 'grandfather rights' and similar *ad hoc* arrangements is going to protect less efficient operators while the alternative of higher landing fees will, again, in the short term, reduce the amount of potential efficiency gain associated with air traffic liberalization which will be passed on in lower fares.

Air traffic control capacity

Air traffic control within Europe is hardly state of the art and, as many air travellers are only too willing to testify, severe congestion problems arise in particular sectors both during certain times of the day and during the summer, holiday season. Some of the difficulties stem from industrial disputes but the system itself is in need of updating. The problems are also compounded because of the complexities which stem from the amount of air space reserved for military use. Because of this the average aircraft flies 10% further than the most direct route and some routes, such as that between Brussels and Zurich, are 45% longer than the minimum path (Commission of the European Communities, 1989). Further, the pattern of demand for air travel in Europe, at least under the regulatory system operational in the late 1980s, tends to lead to a concentration of traffic on certain routes (see again Figure 4.2) with resultant problems of congestion.

Liberalization of aviation markets in the USA brought forth considerable investment in new air traffic control infrastructure, e.g. some $4 billion was spent on research and investment in automation alone. The systems operational in Europe are, in general, dated and in need of replacement. This is particularly true in some of the central areas of the Continent where air traffic routes converge and traffic tends to be concentrated. The problem is recognized but planning of new infrastructure is slow.

Air traffic control in Europe has traditionally been a national concern but with a high degree of co-ordination among some nations (through Eurocontrol) being an inevitable consequence. There are a number of Air Traffic Flow Management Units (ATFMUs) and traffic is passed between them. Direct communications links between them are, however, poor with only Frankfurt, Paris, London, Madrid and Rome linked by a telecommunications system. It is planned by the mid-1990s that a Central Flow Management Unit (CFMU) will become operational so that an overview of flights can be obtained. Further, a Future European Air Traffic Management System (FEATS) is proposed for about the year 2010 which would embody co-ordinated technical planning.

Actual investment is, though, much slower in coming. The UK has a five-year programme, costing some £0.6 billion, but this is perhaps the firmest plan for investment. Recent reports indicate that there are tentative plans for a further £0.6 billion to be spent by other European nations but this is not definitely committed.

4.6 EC liberalization

Europe obviously extends beyond the EC and, as we have seen above, there are several other international bodies with an interest in European aviation. Nevertheless, the EC does represent a large aviation market and its attitude towards liberalization has important implications for other European states. Indeed, the increasing number of mergers and links between EC airlines and other European carriers makes this inevitable. Further, changes have been taking place in the EC's attitude towards aviation which are of intrinsic interest in themselves.

A considerable part of the pressure for reform of European aviation policy stems from the general movement towards a freer market within the EC. In 1957 the Treaty of Rome set out to establish economic integration in the European Economic Community (EEC) by a process of competitive trade interpenetration. It set out general conditions designed to prevent trade distortions within the Community and to ensure freedom to supply services. Article 3(f) of the Treaty, for example, declares that the Community shall seek to establish conditions which ensure that competition is not distorted. The competitive thrust in the case of goods was to be achieved by the elimination, harmonization, etc of: internal tariffs, quantitative restrictions and a variety of non-tarriff barriers. The Treaty also provided for freedom to supply services and for the free movement of factors of production (see Swann, 1988b).

Further, because of the supposed special conditions prevailing in transport, Article 3 of the Treaty also provided for a separate regime for transport – the Common Transport Policy (CTP). While details of the intended form of the CTP were limited, the provisions which were set down were liberalizing in their nature. However, they were only to apply to road, rail and inland waterways. Decisions about air and maritime transport were deferred until such time as unanimous agreement was reached by the Council of Ministers.

Aviation was, therefore, not automatically included in the CTP. Indeed, referring back to our earlier examination of European aviation in the 30 years since the signing of the treaty there are clear indications that some EC members have exploited this situation and often used it to resist pressures for liberalization. Bilateral agreements between members, for example, have regulated capacity, controlled fares and resulted in pooling agreements. Also, foreign investment in national airlines has been excluded or limited in most countries.

Although we have argued that European aviation has possibly been more competitive than is sometimes thought, the key point is that distortions to competition have been manifest and much of the regulation of aviation has been contrary to the spirit of the Treaty of Rome. By the mid-1980s this was appearing out of line with wider developments along more liberal lines within the Community. An interesting question, however, is why this situation was allowed to continue for so long (Button & Swann, 1989a). This can be looked at in two contexts.

First there is the issue of Community anti-trust policy which covers such things as inter-state collusionary arrangements and abuses of market dominating positions, and the granting of special rights to public enterprises (see Argyris, 1989). Why has this not been applied to aviation? In the first place there has been uncertainty over the position of air transport under the Rome Treaty. Certain articles in the Treaty state that the CTP only applies to road, rail and inland waterways and that the Council of Ministers should decide on the provisions made for air transport (Article 84). The question then arises, however, as to whether the remainder of the Treaty applies to aviation. A succession of judgements by the European Court of Justice between 1974 and 1978 made it clear that other articles relating to such things as competition (Articles 85, 86 and 90), state aid, etc *did* apply to aviation. This did not end the matter and the Commission defended its inertia for a number of years on the basis that fare-setting activities were autonomous actions of government, and not of enterprises, and thus outside of the rules governing competition. Further, the Commission was in the rather difficult position that the Council of Ministers did not confer implementing powers on the Commission with regard to aviation and hence the latter could only investigate infringements of the competition rules in collaboration with the member states. The Commission could not impose penalties itself.

With regard to designation, there are two points to be made. First, the Treaty prevented discrimination in establishment and freedom to supply on grounds of nationality. In many cases, however, it is difficult to prove that this was occurring and national regulators could point to their own airlines being refused licences. Second, the rules (Article 90) pertaining to competition and state aid in the context of public undertakings and enterprises to which the state grants exclusive rights have caveats attached to them. Essentially, the Treaty rules only apply to them if they do not prevent the enterprises supplying services of general economic interest from attaining the tasks assigned them. Airlines, it has been claimed, following precedents set in other sectors such as broadcasting, do not fall into this category (Kuijper, 1983).

The Commission was not, however, entirely inactive in the 1970s when it came to developing an aviation policy. It suggested reforms as early as 1972 (*Official Journal of the European Communities,* 1972) but the first significant proposal, known as *Civil Aviation Memorandum Number 1* (Commission of the European Communities, 1979) came seven years later. This proposed:

● Increased possibilities for market entry and innovation were desirable but full freedom of access was a long-term prospect;
● There was a need for the introduction of various forms of cheap fare;
● There was a need to develop new cross-frontier services connecting regional centres within the Community;
● An implementing regulation applying the competition rules directly to air services was essential;
● Increased competition emphasized the need for a policy on state aid to airlines;
● While the right of establishment applied directly to airlines, Council action was necessary since practical and political obstacles would otherwise exist.

Movement, however, was slow. The early 1980s, however, partly as the result of the individual actions of Lord Bethnell (a UK Member of the European Parliament who albeit unsuccessfully, took the Commission to the European Court of Justice over regulatory controls) saw mounting pressure for change. The Commission of the European Communities (1981) began gathering information regarding scheduled air fares and was subsequently encouraged to probe further by the Council of Ministers.

In 1983, though, the Council of Ministers adopted a directive for the development of scheduled inter-regional services. This reduced the grounds upon which licences could be refused and, in fact, as Pelkmans (1986) puts it, the situation became one of near-automaticity. The caveat to this was that the types of service to which it applied (e.g. kind of aircraft, length of route, etc) rendered it largely ineffective. Modified proposals extending the range of regional services which would fall within the directive were proposed by the Commission in 1989.

In 1984 came the Commission's *Civil Aviation Memorandum Number 2* (Commission of the European Communities, 1984), which moved the 1979 statements of desirability onto more liberalizing proposals. The new elements compared to the 1979 document were:

- Fares should be subject to a zone of flexibility. A reference fare level and a zone of reasonableness should be arrived at for international services on the basis of official double approval by both states involved. Having established the scope for flexibility, air fares within the zone of flexibility should be the subject of country of origin approval or double disapproval.
- The dominance of national flag fliers within the bilateral system was to remain relatively undistured, except for a minor change to the effect that other airlines would be allowed to enter and take up any unused route operating rights.
- In the case of 50–50 divisions of traffic between the flag carriers of each state, state A could not oppose a build-up of traffic by state B until A's share had fallen to 25%.
- Inter-airline agreements should be subjected to control. Capacity agreements would be permissible, provided airlines were free to withdraw. Revenue-sharing pools might also be exempted provided their revenue redistributing effect was minimal.

Despite the initiatives of the Commission and efforts by several Members of the European Parliament, the real impetus for substantive change came in 1985 with the *'Nouvelles Frontières'* case (European Court of Justice, 1986). In this judgement the European Court of Justice ruled on the activities of a French travel agent's air fare cutting activities. The relevance of this was that it indicated to the Commission that its powers to attack fare setting in aviation are somewhat stronger than it thought. Essentially, the Court said that if the Commission or an appropriate national authority was to pronounce adversely on an airline restriction then national courts would have to take note of this. In other words a party 'with standing' could raise the issue of restrictions in a national court and the agreement might fall. The Commission, being somewhat concerned about the rather slow pace of the reforms being pursued by the Council of Ministers began instigating proceedings against airlines on this basis in 1987 (Commission of the European Communities, 1987b). This offers illustration of the way in which the Commission is able, on occasions, to take advantage of events to force the Council to act.

The response of the Ministers (which also subsequently resulted in the withdrawal of legal actions by the Commission) was to introduce a slightly more radical packages of measures in 1987 (see Argyris, 1989; Sorenson, 1990). A permanent measure allowing the Commission to apply rules relating to competition directly to aviation (excluding intra-state and operations to third countries) was adopted. Enabling legislation in 1988, however, does allow the Commission to exempt for a limited duration (to expire by 1991) *en bloc* three categories of agreement:

- concerning joint planning and co-ordination of capacity, revenue sharing, consultations on tariffs and aircraft parking slot allocation;

- relating to CRSs;
- about ground handling services.

Whether these exemptions are likely to lead to airlines pursuing anti-competitive activities in practice is difficult to say. In some areas it does seem unlikely. ECAC and the Commission have already drawn up codes of conduct to reduce the most blatant forms of abuse of CRS systems – although these differ to that employed in the USA. Further, as we have pointed out above, the form of ownership of the systems being used in Europe is different to that found in the USA and the ability to abuse the system less. Indeed, even under the bloc exemption, should vendors abuse their position the exemption could be revoked and, ultimately, the Council could divest the systems. The other exemptions, however, may prove more serious.

The Council also adopted a directive in 1987 designed to provide airlines with greater pricing freedom. The enabling legislation set out above allows for collusion in the short term but individual airlines can act individually when applying for approval of air fares. A state authority is directed to accepted a fare application if it is reasonably related to economic costs, i.e. it cannot reject a fare simply because it undercuts another operator. The Civil Aviation Authority in the UK, for instance, cannot reject a fare proposal from Air France on the London/Paris route simply because it is lower than British Airways' fare. Disputes over a fare go to arbitration. While the fare-setting procedure applies primarily to economy scheduled fares, the directive also provides scope for discounts. Provided certain conditions are met, fares can be reduced by varying amounts. Fifth freedom operators (i.e. third party airlines carrying passengers between two other states) are permitted to match any discounts.

A phased policy of liberalizing market access over three years was also initiated by a decision of Council in 1987. The traditional 50–50 division of traffic common in most bilateral pooling agreements was to be relaxed so that, in the period to October 1989, where competition existed the share could change up to a 55–45 split and state authorities were not permitted to intervene. The split could become as wide as 60–40 thereafter. This means that countries such as Italy, Greece, Denmark and France which have traditionally rigidly controlled capacity will have to liberalize to a considerably extent.

Further, over a period of time member states gradual lose the right to refuse multiple designation by other member states on a city-pair basis widening the scope of national authorities to designate more than one carrier to each route. Multiple designation is allowed on a country-pair basis. Automatic fifth freedom rights were also extended but effective competition from such activities was constrained by specific stipulations about the nature of services which could be offered and the types of airport which must be used. Additional to all this, the Commission reminded member states of their obligation to open ownership of air carriers to nationals of other member states.

Regulation of European aviation is a continuing process and the trend is towards further liberalization. While the new package of measures (now commonly called phase one) were being implemented within the EC the Commission developed proposals for the next phase of policy reform. As a pragmatic measure, during a second phase of liberalization the EC Commission proposed in 1989 the encouragement of greater use of regional airports as a priority. This was to meet the high levels of forecast demand for air travel at a time when there are severe limits on infrastructure capacity. There would also need to be actions to encourage the use of larger aircraft and, *ipso facto,* reduced air movements for a given volume of passengers. The question of the appropriate mechanism for slot allocation in this type of environment is a crucial one and the EC Commission seems uncertain on the way forward. In the short term it is still, for example, developing strategies which could involve such things as reforms in landing slot charges, the introduction of slot auctions at airports, frequency capping the use of small aircraft, etc, in order to make better use of existing infrastructure until new capacity comes on-line (Sorenson, 1990).

Late 1989 saw a rather sudden and slightly unexpected movement forward in the liberalization process. An accord (a firm political agreement but still requiring facilitating legislation by national governments) was reached designed to move towards a single European Market by the beginning of 1993. In part the rather sudden change may have been influenced by lobbying in Brussels on behalf of the French airline UTA which was complaining about unfair discrimination against it in the alloction of European licences. As mentioned earlier UTA was refused European route licences by the French government on a number of routes currently served by the state-owned Air France.

It was agreed by the Council in 1989 that by January 1993 the following would be completed:

● The removal of government to government capacity-sharing arrangements guaranteeing each airline a certain percentage of the market.
● Governments accepted the principle of double disapproval for fares. This would remove the ability of just one country in a bilateral agreement to veto a change in airline fares – both would have to agree to block such a change. The measure would be introduced in a staggered fashion with fare zones being introduced allowing airlines to cut there prices to as low as 30% of the standard economy tariff subject to the double disapproval rule until by 1993 they will be free to set the fares they wish subject to the rule.
● Governments would cease to discriminate against airlines provided that they met technical and safety standards and ran economically.

In addition to this accord, at the same meeting of the European Community Transport Ministers, agreement was reached on ownership rules. Until then an airline had to prove that it was 'substantially owned' by one European nation before it was allowed to fly from that country. The new rules abolish this rule over a two-year period. This means that fifth rights become automatic within the EC.

It clear that changes are continuing to take place in European aviation policy but the process of liberalization has been slow. Some controls remain (e.g. over cabotage, qualifications of pilots, air traffic controllers, etc) and are likely to do so for some time to come (McGowan & Trengrove, 1986). The airline industry itself, however, is a very adaptable and flexible one. This adaptability pulls in two directions. In the past it has sometimes allowed for greater competition and market access than the regulatory regime would imply (e.g. witness the growth in non-scheduled operations). But on the other hand, it may result in airlines circumventing rules designed, for example, to prevent anti-competitive practices. This latter is returned to, in a wider context, below.

4.7 Global considerations

European aviation is part of a much larger aviation market and as such it is influenced by changes in world aviation but also, because of the importance of both its air carriers outside Europe and of the European market itself, influences these changes. First it is useful to consider the implications for and reactions of European airlines both to the changes taking place in Europe and to the developments in wider aviation markets.

There are already signs that airlines are adapting their stances in the face of market trends and new regulatory regimes. This is particularly so over the question of size. While we have seen that the US experience is not entirely a satisfactory basis to predict future events in Europe, there does seem to be some evidence that size is important in a more liberalized market place. There are, for instance, already indications that European airlines are beginning to seek ways of expanding in efforts to maximize economies of scope and density. Some have taken the classic route of direct takeovers (e.g. British Airways' takeover of British Caledonian and Air France's efforts to takeover UTA being the most dramatic examples) but in other cases airlines have attempted to strengthen their positions in other ways. Within Europe, British Airways and KLM have each bought a 20% share in the Belgium carrier Sabena (although at time of writing, formal objections were being lodged by other carriers in the UK against this) while Swissair has attempted to strengthen its position by going outside Europe and has swapped 5% shareholdings with the US carrier Delta Air Lines and SAS has a 10% equity holding in Texas Air and has made alliances with All Nipon Airways, Thai International, Lan-Chile and Canadian Airlines International. KLM is also buying into Northwest. At a slightly different level British Airways and United Airlines of the USA agreed in 1987 to integrate their flight schedules and networks in the USA with a number of joint services being offered. Attempts at closer links, however, with British Airways being involved as a member of a group trying to take over United, fell through in the autumn of 1989 due to problems in financing the deal.

Linked to the above, there is the wider international market. This ties in with the question of scale and the prospect of international mega-carriers dominating the market through extended hub-and-spoke operations. There are already indications (see Table 4.9) that the largest US carriers are extending their international operations following both domestic deregulation and the general liberalization of international markets. In particular, there are signs (Table 4.10) that domestic deregulation lowered US airline costs and provided them with the prospect of more effectively competing in international markets with airlines from other countries, coupled with this, the US airlines' system or hubs, and the cost-efficient services radiating from them, meant that carriers could afford to offer almost 'loss-leader' fares to attract overseas travellers onto their domestic network.

This expansionary trend on the part of US aviation gives added impetus to the process of takeovers, mergers and agreements which are being experienced with the European market. It also raises broader questions about cabotage and forms of operation within the USA and the EC in particular. Code-sharing agreements and multi-national ownership of airlines makes the definition of national and intra-Community operations rather an arbitrary one. One can, therefore, anticipate a long-term trend not just to larger airlines but also to horsetrading which will produce agreements on intra-EC cabotage rights for outside carriers.

Table 4.9 Foreign gateways served by selected US carriers.

Carrier	1978	1986
Delta	1	5
American	0	8
United	0	14
Texas Air	3	21
Total	4	48

Source: James, G. (1988) *Overview: Pre- and Post-deregulation US Airline System*, Presentation to the Transportation Research Board Annual Meeting, Washington.

Table 4.10 Average annual percentage decline in unit costs and sources of unit costs for US and non-US carriers pre- and post-deregulation.

Sources	Pre-deregulation (1970–75)		Post-deregulation (1975–83)	
	USA	Non-USA	USA	Non-USA
Productive efficiency				
Operating characteristics	1.6	3.3	2.2	2.4
Technical efficiency	1.4	1.2	1.1	0.4
Total productive efficiency	3.0	4.5	3.3	2.8

Source: Caves, D. W., Christensen, L. R., Tretheway, M. W. and Windle, R. J. (1987), 'An Assessment of the Efficiency Effects of US Airline Deregulation via an International Comparison', In Bailey, E. E. (ed), *Public Regulation: New Perspectives on Institutions and Policies*, Cambridge Mass: MIT Press.

Externally, the globalization of air transport can pose other problems. At present there is only limited incentive for flag carriers to merge because international services between EC members and other countries are determined by bilateral agreements. Bilaterals often require the designated airlines to be owned and controlled by nationals of the countries involved. Consequently, say in the case of the US–Italy bilateral then if Air France acquired Alitalia then the bilateral would be undermined – the now French-owned airline would be relegated to at best offering fifth freedom services. Community status for airlines, if this were ever acceptable to outside countries, could be a means of circumventing this type of problem in the longer term.

4.8 Some concluding comments

In subsequent years the 1980s may be seen as the age of regulatory reform. Market forces were freed in many countries of the world as governments withdrew from economic regulation. One can point to the major reforms in Eastern bloc countries during the latter part of the decade as perhaps the most dramatic and surprising of these reforms. Many markets in Western economies, however, have also been the subject for considerable liberalization. Transport markets have, perhaps, been at the forefront of these latter changes.

Towards the end of the decade, within Europe the progress towards a Common Transport Policy, with a strong market liberalizing orientation, has gathered pace. Aviation is now part of this policy. In particular, there have been important moves to develop a structure of aviation regulation which will permit a high level of effective competition within the industry. What is equally clear, however, is that the existing system differs in many important ways from the open market approach which has been adopted in the United States and, further, it seems clear that the American approach will never be fully embraced within Europe.

Further, it does not mean problems do not remain within the framework which has emerged. Questions of cabotage rights and their role as links between domestic and intra-European international services are an obvious example. The long-term approach to mergers policy is equally still not completely clarified. The Commission seeks a way of ensuring that the European market does not become dominated by mega-carriers but at the same time ensuring that in the larger international market European airlines are not disadvantaged by their size from competing efficiently with US and other major carriers. Fostering a policy of allowing mergers where there exists complementarity but preventing those which reduce competition is, however, difficult to put into practice on a consistent basis. The situation is compounded by the expansion of such things as marketing agreements, minority shareholdings of one airline in another and the growing practice of backing shares by equity capital. The hand of the Commission was strengthened in this respect in late 1989 when it was agreed that the EC mergers policy would embrace laws requiring that mergers involving undertakings with annual global turnovers of more than ECA 2 billion (with supplementary limits for firms mainly trading within the EC) would be called in by the Commission. In other words a proactive element was introduced into competition policy. Indeed, the proposed merger between Air France and UTA was one of the first to come within the new code.

The process towards liberalizing intra-EC aviation has unquestionably been gradual but there have, nevertheless been significant steps forward over the past decade and further advances seem inevitable. The need to deal with 12 countries has made the opportunities for rapid and planned trasformation along the lines of the USA impossible. Such speed may also have been undesirable for the EC in the sense that the costs of being first in the field were borne by US carriers and travelling public, while those concerned with EC policy have had the opportunity to wait and observe before acting. There is already evidence that this has happened with respect to such things as CRS controls and attitudes towards mergers policy. The danger with this approach is that policy makers may wait too long and in seeking to ensure optimality in decision-making actually miss important opportunities. Further, there are aviation markets outside the EC where the lack of a clear policy within the Community, coupled with the prospect of EC airlines not being fully attuned to the full rigours of competitive and contestable market conditions, could, in the long term, adversely affect the performance of European carriers in the world market. In particular, it could give US and Asian carriers greater scope to expand their international operations at the expense of European airlines.

References

Argyris, N. (1989) 'The EEC rules of competition and the air transport sector', *Common Market Law Review*, **26**, 5–32.

Ashworth, M. & Forsyth, P. (1984) *Civil aviation policy and the privatization of British Airways*, London: Institute of Fiscal Studies.

Association of European Airlines (1984) *Comparison of air transport in Europe and the USA*, Brussels: AEA.

Bailey, E. E. (1985) 'Airline deregulation in the United States: the benefits provided and the lessons learned', *International Journal of Transport Economics*, **12**, 119–44.

Bailey, E. E. & Panzar, J. C. (1981) 'The contestability of airline markets during the transition to deregulation', *Journal of Law and Economics*, **44**, 125–45.

Barrett, S. (1987) *Flying high, airline prices and European regulation*, Aldershot: Avebury.

Baumol, W. J., Panzar, J. C. & Willig, R. D. (1982) *Contestable markets and the theory of industrial structure*, New York: Harcourt Brace Jovanovich.

Button, K. J. & Gillingwater, D. (1986) *Future transport policy*, London: Routledge.

Button, K. J. (1989a) 'The deregulation of US interstate aviation: an assessment of causes and consequences (Part 2)', *Transport Reviews*, **9**, 189–215.

Button, K. J. (1989b) 'Infrastructure plans for Europe', *CERUM* Working Paper, CWP-1989:10, Umea: Umea University.

Button, K. J. & Swann, D. (eds) (1989a) 'European Community airlines – deregulation and its problems', *Journal of Common Market Studies*, **27**, 259–82.

Button, K. J. & Swann, D. (1989b) *The age of regulatory reform*, Oxford: Clarendon.

Civil Aviation Authority (1983) *A comparison between European and United States fares*, CAA Paper 83006, London: CAA.

Civil Aviation Authority (1984a) *Deregulation of air transport*, CAA Paper 84009, London: CAA.

Civil Aviation Authority (1984b) *Airline competition policy*, CAP 500, London: CAA.

Civil Aviation Authority (1987) *Competition on the main domestic trunk routes*, CAA Paper 87005, London: CAA.

Civil Aviation Authority (1988) *Statement of policies on air transport licensing – June 1988*, CAP 539, London: CAA.

Commission of the European Communities (1979) 'Contribution of the European Communities to the development of air transport services', *Bulletin of the European Communities*, Supplement 5.

Commission of the European Communities (1981) *Report on scheduled passenger air fares in the European Economic Community*, COM (81) 398 Final, Brussels: EC.

Commission of the European Communities (1984) *Civil aviation memorandum No. 2, towards the development of a Community air transport policy*, COM (84) 72 Final, Brussels: EC.

Commission of the European Communities (1985) *Completing the Common Market*, COM (85) 310 Final, Brussels: EC.

Commission of the European Communities (1987a) *The likely impact of deregulation on industrial structures and competition in the Community: final report*, Brussels: EC

Commission of the European Communities (1987b) *Sixteenth report on competition policy*, Brussels: EC.

Commission of the European Communities (1989) *Air traffic system capacity problems*, COM (88) 577 Final, Brussels: EC.

Doganis, R. (1985) *Flying off course: the economics of international airlines*, London: George Allen & Unwin.

Erdmenger, J. (1983) *The European Community transport policy: towards a common transport policy*, Aldershot: Gower.

European Civil Aviation Commission (1981) *Report of the task force on competition in intra-European air services*, Paris: ECAC.

European Court of Justice (1986) *Ministère Public v Lucas Asjes and Others*, Cases 209–213/84.

Forsyth, P., Hill, R. & Trengrove, C. (1986) 'Measuring airline efficiency', *Fiscal Studies*, **7**, 61–81.

Gomez-Ibanez, J. A. & Morgan, I. P. (1984) 'Deregulating international markets: the examples of aviation and ocean shipping', *Yale Journal on Regulation*, **2**, 107–44.

House of Commons Transport Committee (1988) *Airline competition: computer reservation systems*, HC 461, London: HMSO.

House of Lords Select Committee on the European Communities (1980) *European air fares*, HL 235, London: HMSO.

International Air Transport Association (1974) *Agreeing routes and fares*, Montreal: IATA.

Kahn, A. E. (1988) 'Surprises of airline deregulation', *American Economic Review, Papers and Proceedings*, **78**, 316–22.

Kasper, D. M. (1988) *Deregulation and globalization: liberalizing international trade in air services*, Cambridge, Mass: American Enterprise Institute/Ballinger.

Keeler, T. E. (1984) 'Theories of regulation and the deregulation movement', *Public Choice*, 44, 103–45.

Keeler, T. E. (1990) 'Airline deregulation and market performance: the economic basis for regulatory reform and lessons from the US experience'. In Banister, D. & Button, K. J. (eds) *Transport in a free market economy*, London: Macmillan.

Kuijper, P. J. (1983) 'Airline fare-fixing and competition: an English Lord, Commission proposals and US parallels', *Common Market Law Review*, 20, 203–32.

Labich, K. (1989) 'Should airlines be reregulated?' *Fortune*, 119, 78–82.

Levine, M. (1987) 'Airline competition in deregulated markets: theory, firm strategy, and public policy', *Yale Journal on Regulation*, 29, 393–494.

Lissitzyn, O. (1968) 'Freedom of the air: scheduled and unscheduled services'. In McWhinney, E. & Bradley, M. (eds) *The freedom of the air*, New York: Oceana.

McGowan, F. & Trengrove, C. (1986) *European aviation: a common market?* London: Institute for Fiscal Studies.

Morrison, S. (1989) 'US domestic aviation'. In Button, K. J. & Swann, D. (eds) *The age of regulatory reform*, Oxford: Clarendon.

National Consumer Council (1986) *Air transport and the consumer*, London: HMSO.

Official Journal of the European Communities (1972) C 110, 18 October.

Organization for Economic Co-operation and Development (1979) *Competition policy in regulated sectors*, Paris: Organization for Economic Co-operation and Development.

Organization for Economic Co-operation and Development (1988) *Deregulation and airline competition*, Paris: Organization for Economic Co-operation and Development.

Pelkmans, J. (1986) 'Deregulation of European air transport'. In de Jong, H. W. & Shepherd, W. G. (eds) *Mainstreams in industrial organization*, Dordrecht: Martinus Nijhoff.

Pelkmans, J. (1990) 'Deregulation in European air transport: issues after 1992'. In Banister, D. & Button, K. J. (eds) *Transport in a free market economy*, London: Macmillan.

Pryke, R. (1987) *Competition among international airlines*, Thames Essay 46, London: Trade Policy Research Centre.

Pryke, R. (1988) 'European air transport liberalization', *Travel and Tourism Analyst*, 6, 5–18.

Pryke, R. (1990) 'American deregulation and European liberalization'. In Banister, D. & Button, K. J. (eds) *Transport in a free market economy*, London: Macmillan.

Sawers, D. (1987) *Competition in the air: what Europe can learn from the USA*, Research Monograph 41, London: Institute of Economic Affairs.

Seidenfuss, H. (1983) 'The interface between air transport and land transport in Europe', paper presented to a meeting of the European Conference of Ministers of Transport, Paris.

Sorensen, F. (1990) 'The changing aviation scene in Europe'. In Banister, D. & Button, K. J. (eds) *Transport in a free market economy*, London: Macmillan.

Stigler, G. J. (1971) 'The economic theory of regulation', *Bell Journal of Economics and Management Science*, 2, 3–21.

Swann, D. (1988a) *The retreat of the state*, Hemel Hempstead: Harvester Wheatsheaf.

Swann, D. (1988b) *The economics of the Common Market*, 6th ed., Harmondsworth: Penguin.

Tucci, G. (1985) 'Regulation and "contestability" in formulating an air transport policy for the European Community', *Rivista di Politica Economica*, 19, 3–23.

CHAPTER 5

Airline deregulation in Canada

Tae Oum, William Stanbury and Michael Tretheway

5.1 Introduction

Although Canada has the second largest land mass in the world, its population is not yet 26 million. Air transportation is a vital link in unifying the nation and conducting its trade. Because of its importance to both the economy and the social fabric of the nation, air transport has been the focus of much public policymaking. As early as 1919, Canada was developing policies toward the industry. By the end of the 1930s it had established a government-owned airline to provide transcontinental services, and had established a comprehensive regulatory regime to control entry and prices. In every decade since, major air transport policy issues were debated and resolved.

As the US moved to deregulate its airline industry in the late 1970s, forces were set in motion to do the same in Canada, although they were not then seen as steps toward total deregulation. The example and experience of the US could not be ignored, given the common border, common language (for 75% of the population), and the extent of the penetration of US communications media in Canada. Because of the proximity of the border to most major Canadian population centres, Canadians were able to vote with their dollars and patronize US rather than Canadian carriers by crossing the border. The pressures were clear and changes were made in Canadian policy and regulation. However, unlike the US which made a rapid change from a regulatory to a deregulated regime, Canada's approach was more gradual: evolutionary rather than revolutionary. Some freedom to offer low fares was granted in 1979. Much more freedom was given in 1984. It was not until 1 January 1988, however, that deregulation became official. Even then, the sparsely populated northern region of Canada continued to operate in a regulated environment, although that environment is substantially less restrictive than the previous one.

This chapter discusses regulatory reform and deregulation in Canada. It starts by discussing the regulatory regime in place in the late 1970s (section 5.2), the forces behind deregulation (section 5.3), and Canada's evolution toward deregulation (section 5.4). Following this, the response of the industry is studied. Section 5.5 describes the significant changes in industry structure, while section 5.6 gives data on the early experience with conduct and performance. A critical, although recent, change in structure is the

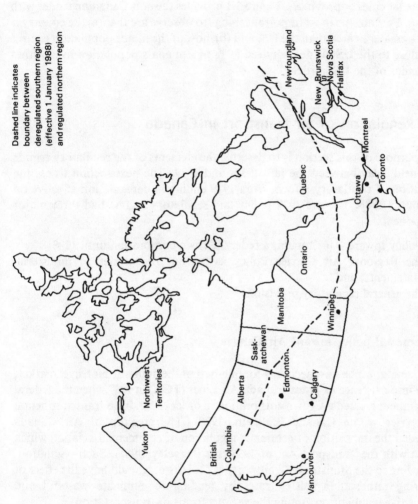

Dashed line indicates
boundary between
deregulated southern region
(effective 1 January 1988)
and regulated northern region

Figure 5.1 Map of Canada.

privatisation of Air Canada, and this is treated in a separate section (section 5.7). Recognizing that the industry is still in transition, section 5.8 speculates on future developments. Finally, section 5.9 attempts to draw some conclusions on the Canadian experience with airline regulatory reform. These are presented in the form of lessons for other countries contemplating their own reform.

In the discussion which follows, it is inevitable that reference will be made to particular cities or provinces. Figure 5.1 provides the non-Canadian reader with a map. Perhaps the most important things to observe are that (a) the country is vast – greater in area than the US, and (b) most of the major population centres are close to the US border – indeed 80% of Canada's population lives within 200 miles of the US border.

5.2 Regulation of air transport in Canada

The purpose of this section is to describe the elements of the regulatory regime that held sway between the late 1930s and 1979. The next section traces the evolution of regulatory reform from 1979 ending in deregulation effective on 1 January 1988. The pre-1979 policy and regulatory regime had three major components:

(1) policy toward Air Canada, a federal Crown corporation until 1988;
(2) the Regional Air Carrier Policy as set out in a series of ministerial statements; and
(3) the general regulatory provisions.[1]

Government policy toward Air Canada

Air Canada, which now accounts for one-half of the domestic air travel market, came into existence as Trans-Canada Air Lines (TCA) in 1937 when the federal government created a Crown corporation in order to establish transcontinental air service *within* Canada (Baldwin, 1975). The creation of Air Canada preceded the imposition of economic regulation of commercial aviation which began with the Transport Act of 1938 (see Baggaley, 1981). Such regulation, according to the Minister of Transport, C. D. Howe, 'should have the effect of lowering rather than raising rates by tending to eliminate wasteful and destructive competition among the various forms of transportation.'[2]

It was government policy between 1937 and 1959 that Air Canada have a monopoly on all domestic transcontinental routes. For example, in 1943 Prime Minister Mackenzie King told Parliament that competition over the same route would not be permitted whether between Air Canada (then TCA) and a privately-owned carrier or between privately-owned carriers.[3]

Between 1937 and 1978 the federal Cabinet approved the routes and fares proposed by Air Canada and the regulators were required to issue the necessary licences or approve Air Canada's fares as set by the federal Cabinet.[4] In 1944 the government made it clear that its contract with Air Canada took precedence over the requirements of the regulators. The purpose of this policy was to eliminate the need for direct cash subsidies to extend air service to smaller centres. With the Air Canada Act of 1978, the Crown carrier was finally made subject to the same regulatory regime as all other carriers – see Gillen (1979).

Although the Conservative government allowed CP Air to provide transcontinental service beginning in 1959, the Cabinet limited it to one flight per day each way between Vancouver and Montreal in order to protect Air Canada's financial position. Between 1959 and 1965 CP Air obtained an average of 12.7% of the transcontinental market (Baldwin, 1975: 149). Capacity restraints were gradually relaxed pursuant to Ministerial statements in 1967 (2 flights per day and 25% of capacity by 1970), 1974, and 1977 (capacity increased to 35% of the growth in 1978 to 45% in 1979). In 1979 all capacity constraints on CP Air were removed.

Under the Regional Air Carrier Policy (see below) Air Canada was expected (as was Canadian Pacific Airline Ltd – CPAL) to give up shorter, thinner regional routes to assist the five regional carriers.[5] In addition the national carriers were not to compete with the regionals on their more important, larger, intra-regional routes. The policy, however, was designed to benefit Air Canada as well as to protect the regionals.

Air Canada had a monopoly on all transborder routes until 1967 when CP Air obtained Vancouver–San Francisco.[6] Since that time, the federal government has given Air Canada the lion's share of transborder routes negotiated with the US. For example, in 1974 Air Canada received 14 of the 17 new transborder routes (CP Air, PWA and Nordair each obtained one). CAIL still has only four US points: San Francisco, Los Angeles, Seattle and Chicago (Midway).

From 1937 to 1948 it was government policy for Air Canada to have a complete monopoly on international routes. In 1948, CP Air was named Canada's flag carrier in the Pacific when it was awarded rights to Canada–Australia with stops in Honolulu and Fiji. Air Canada, however, retained all other international routes including the most lucrative North Atlantic routes (eastern Canada to the UK and northern Europe). CP Air's international service expanded in the 1950s and 1960s, although few routes were lucrative.

In 1965 the Minister of Transport announced that the two national carriers had agreed on the areas of the world in which each would be the sole Canadian carrier. CP's area was 'the whole Pacific area, the whole continent of Asia, Australia and New Zealand, Southern and South Eastern Europe and Latin America'. In addition, CP Air would continue to serve Amsterdam. Air Canada was assigned the UK, Western, Northern and Eastern Europe and the Caribbean. The division of the US was left unspecified pending bilateral negotiations. Each airline was to be regarded as the nation's 'chosen instrument' in their own areas. They were to co-operate in selling the services of the other in competition with foreign carriers.

Crown-owned airline Air Canada was favoured in its dealings with the federal government in other ways. For example, it had access to confidential data. Until 1983/84, Air Canada was closely involved in the formation of commercial aviation policy within the Department of Transport. In addition, the airline's top executives had access to the minister and other members of Cabinet. Air Canada was closely consulted on the design and construction of major airports. It was able to obtain its own terminal at Toronto's Pearson airport (Terminal 2) and preferred positions at most others including Mirabel near Montreal.

Further, until 1986, Air Canada obtained the lion's share of all air travel by federal public servants through the vehicle of Central Travel Service. The Crown airline also benefited greatly from the implicit guarantees of its debt by the federal government. For example, in 1978 when it was recapitalized, Air Canada had a debt to equity ratio of 24:1.

Regional air carrier policy

The development of federal policy toward the five regional carriers (many of which were established in the late 1940s and 1950s – see Baldwin, 1975), can be traced to statements by the Minister of Transport on 26 April 1964 and 20 October 1966 (see Baldwin, 1975, ch. 4). However, the most comprehensive policy statement regarding the regionals was issued on 20 October 1966. It established a number of 'principles' under which the regional carriers were to operate.

- Regionals were to operate local or regional routes to *supplement* and not directly compete with the domestic mainline operations of CP Air and Air Canada and to provide scheduled service into the north. They were not to become transcontinental carriers, but rather operate only within their region.
- Routes unsuitable because of their volume of traffic and equipment requirements were to be reserved for regional carriers or, in some cases, transferred to them.
- The national and regional carriers were encouraged to develop joint fare and commission arrangements and co-operate on technical and servicing arrangements, interconnections and even advertising and sales activities.
- A 'limited policy of temporary subsidies' was to be introduced (see Baldwin, 1975: 211).
- Rules regarding the ability of regional carriers to operate domestic charters were to be relaxed. International charters were seen as a 'useful supplement' for the regionals, but such charter work 'must not overshadow domestic operations'.

The last ministerial policy statement concerning the regionals was made on 15 August 1969.[7] It specified the region in which each was to operate:

- Pacific Western Airlines (PWA) – British Columbia and western Alberta;
- Transair – the Prairies and northwestern Ontario (with access to Toronto);
- Nordair – the remainder of Ontario and northwestern Quebec;
- Quebecair – all of Quebec east of Montreal; and
- Eatern Provincial Airlines (EPA) – the Atlantic provinces (with access to Montreal).

Thus the policy that emerged during the 1960s envisaged the development of a small number of regional carriers each to be a 'chosen instrument' to operate local and regional routes as a supplement to the mainline operations of Air Canada and CP Air. Since their role was envisaged as supplementary, it is not surprising that a considerable part of the growth of the regionals resulted from largely voluntary route transfers from the two major carriers (see Greig, 1977). CP Air was pressured by the regulators in 1947 to give up many routes to smaller carriers including PWA.[8] Air Canada was encouraged to transfer routes to various regionals in the 1960s (see Baldwin, 1975, ch. 4).

As the regional carriers grew, they discovered their protected territory was also a box which constrained their ambitions and made them more vulnerable in the era of deregulation. The regionals wanted access to the biggest hub, Toronto, and they wanted to offer at least some inter-regional service. In order to ensure that its purchase of control of Transair in 1977 would be an economic success, PWA had to make a deal with Air Canada. The Crown carrier agreed to drop certain points in Saskatchewan so the two regionals' routes could be linked, but required Transair to drop its points east of Winnipeg to Toronto. PWA did not regain access to Toronto until 1984 when it was permitted by the CTC to fly Calgary–Brandon–Toronto. EPA wanted access to Toronto (its previous access to Montreal was acknowledged in the 1969 policy statement), and finally got it in 1980 when the Cabinet greatly changed the CTC's decision to allow CP Air to serve Halifax–Toronto non-stop.

General regulatory provisions

In this section the key elements of Canada's system of airline regulation as it existed in the late 1970s and early 1980s are identified and described. In its most important respects, the system was similar to 1938 when economic regulation was first imposed.

Complete control over entry into the industry

Under the Aeronautics Act the Air Transport Committee (ATC) of the Canadian Transport Commission was required to determine if the new carrier's services were required by 'present and future public convenience and necessity' (PCN). The onus was at all time on the applicants to show that there was an unambiguous need for the services they intended to provide, and that this need was not now being met by existing carriers nor were they likely to offer such services. Most importantly, the ATC required the applicant to show that his entry would not adversely affect the profitability of existing carriers. In summary, the ATC used its discretion to severely discourage entry into the industry in the name of the stability of the industry.

Complete control over access to routes

Existing carriers were licensed on a route-by-route basis and to serve only those points listed on their licence(s). The same PCN test was applied to route applications as to entry to the industry. In general, the regulators' policy was to carefully control competition so as to ensure the financial success of existing carriers and to provide the financial means to extend the network with regular scheduled (preferably jet) service to more points by means of cross-subsidization. The policy was somewhat asymmetric in that the two national carriers (CP Air, Air Canada) were allowed to drop routes when they would be taken over by a regional carrier. Almost without exception the regulator ratified the deals made by carriers prior to the public hearings.

Extensive regulation of conditions of service

Canadian regulators sought to control competition among carriers and/or to 'manage' even those routes served by a single carrier by attaching restrictive conditions of service to specific route authorities. These included frequency of service, the amount of capacity offered (e.g. size of aircraft) and requirements for intermediate stops.

Complete regulation of fares

It was not until the 1950s that tight regulation of fares was instituted by Canadian regulators. However, beginning with the Transport Act of 1938, carriers were required to both file their tolls and tariffs and to have them approved by the regulator. In the 1950s with the requirement of more standardized accounting practices came the adoption of the distance-related tapered fare formula (see Baldwin, 1975, ch. 7). In the late 1960s, Air Canada's tapered fare formula was effectively adopted by the ATC for application to all airlines and the fares (economy and first class) of all carriers on regularly scheduled flights on the same route were identical. The ATC sought to ensure that price competition did not occur.

In considering requests for across-the-board percentage fare increases, usually by several carriers at once, the ATC focused on past and projected increases in cost, and the financial position of the carriers, although it did not apply anything remotely resembling a rate base rate of return approach to setting fares. Moreover, the ATC made little effort to examine the efficiency of carriers, or the level of fares in the US, although it was pressed to do so in the 1970s and early 1980s by the aggressive intervention of the Consumers' Association of Canada. With the approval of domestic charter class fares in 1979,[9] a move initiated by Air Canada to respond to the Cabinet's decision to allow more domestic advance booking charters, the ATC required elaborate justification before approving each individual promotion, e.g., the 'seat sales' held by most national and regional carriers several times per year in periods of excess capacity.

Detailed regulation of conditions or 'fences' surrounding discount fares
With the exception of the original, full-plane, discount-fare 'Sky Bus' service offered by CP Air in 1979,[10] the ATC always closely regulated the 'fences' or conditions attached to discount fares. The objective was to ensure that price discrimination worked, namely that low fares attracted more price-sensitive visiting friends and relatives (VFR) passengers and diverted very few 'must go' or business travellers from the full fare seats. When discount fares were permitted in some volume in 1979, the following conditions applied: advance booking period, e.g. 30 days; minimum stay period, e.g. seven days; compulsory round trip (no one-way flights); minimum fee for cancellation or change of reservation; and advance payment, e.g. seven days before departure. While some conditions were later relaxed, the ATC tightened up again in 1982 as described in section 5.4.

Regulation of geographical area within which regional carriers may provide service
The regional air carrier policy has been described above.

Merger policy
The control of mergers in the airline industry (or other horizontal agreements) lay with the Canadian Transport Commission which had to determine if a proposed merger would 'lessen competition unduly'. With exception of the Province of Alberta's purchase of PWA in 1974, the CTC never refused to permit any change of control over any airline in levels I, II or III, This included PWA's acquisition of control of Transair in 1978 (admittedly a failing firm), Air Canada's acquisition of control of Nordair in 1979, and CP Air's purchase of EPA in 1984. Moreover, the CTC did not prevent PWA from acquiring 40% of Swiftair (a cargo carrier) in 1982 and 40% of Time Air (the largest third level carrier) in 1984. In other words, the regulator was entirely passive in respect to mergers.[11]

5.3 Forces that pushed the industry toward deregulation

Because the United States deregulated 10 years prior to Canada, this section starts with a review of the economic forces which pushed the US to deregulate its airline industry. Following this, a discussion of the forces within Canada is given.

US forces[12]

There were three primary forces which induced the US to deregulate its airline industry. These are:

(1) the emergence of substitute products for scheduled airline service;
(2) an increase in the power of consumers;
(3) the emergence of concrete information as to how a deregulated industry would actually perform.

All three of these were, of course, inter-related. For example, as substitute products became available consumers were able to exercise greater power over a carrier, and this in turn led to improvement on performance of the substitute services in unregulated markets.

Perhaps one of the most important events which started the march to deregulation was the response of the regulated scheduled airlines to the introduction of wide-bodied aircraft. In the mid 1960s, when these planes (the Boeing 747, Douglas DC-10 and Lockheed L-1011) were on the drawing boards, the manufacturers promised that their lower per seat operating costs would make air transportation available for the first time to a mass market. It was claimed that this technology would enable people to get on a plane and fly to Paris just as they would hop on a bus for a trip to Chicago. However, when these planes actually came into service, airfares did *not* drop. To a large extent, this was due to the regulator (CAB) not entertaining any innovative fares. The airlines responded by flying these high capacity aircraft just as frequently as they had flown the narrow-bodied planes. As a result, load factors fell. By 1971 only 48.5% of the seats on scheduled US airlines had revenue paying passengers in them. Until 1977 the industry never exceeded a 56% load factor.[13] In contrast, a number of chartered carriers such as World and Capital, flew the same wide-bodied aircraft in both domestic and international services. By filling almost every seats, they were able to offer airfares significantly below those of the scheduled airlines. However, in the early 1970s, the charter carriers were only allowed to sell their product to affinity groups. Thus, members of the local sewing club could book a chartered flight from New York to Los Angeles, but the individual traveller could not. Nevertheless, charter carriers were increasingly providing a substitute service to scheduled air transportation. When CAB chairman John Robson lifted the affinity group requirement on charter services in 1976, these carriers were poised to undermine a substantial portion of the scheduled airline market.

Another substitute to the CAB-regulated, scheduled carriers in some states were the relatively unregulated intra-state carriers. Federal airline regulation did not apply to carriers operating within a single state. In three states, California, Texas and Florida, unregulated jet carriers emerged and prospered. For example, in the Los Angeles–San Francisco market, unregulated PSA dominated the market. Important studies by Levine (1965) and Jordan (1970) indicated that the intra-state carriers were offering fares significantly below those of the CAB regulated carriers. The emergence of Southwest Airlines in Texas had an especially interesting lesson for the industry. When Southwest entered the Dallas–Houston–San Antonio market, the total traffic base expanded considerably. Southwest was *not* stealing traffic from CAB regulated carriers. Its low fares were instead stimulating a wholly new air transport market segment.

This lesson of how low fares stimulate the entire market was not lost on the scheduled airlines. In response to requests by scheduled airlines, Boeing developed the concept of airline seat management.[14] This is a pricing system which allows carriers to maintain high fare revenues from existing customers, while stimulating the market with low airfares targeted at individuals who otherwise would not fly. A critical aspect of seat management is the ability to price discriminate and to enforce this price discrimination.

By the middle of the 1970s, consumers were voting with their dollars by patronizing the substitute charter and intra-state services. The growing evidence of these unregulated carriers successfully providing air transportation at low prices led to significant consumer pressure on politicians to reform airline regulation. By 1976, both Democrats and Republicans were supporting deregulation. Studies of intra-state and charter carriers, couple with the recent success of unregulated commuter airlines and Federal Express, were providing consumers, carriers and politicians with concrete information as to how a deregulated industry might perform.[15]

By 1976/77, deregulation was inevitable. The 1976 removal of affinity requirements on the charter carriers and the 1977 introduction of low fare service by Laker Airways between North America and Europe threatened a major diversion of traffic from the scheduled airlines. The scheduled airlines asked for and received authority in 1977 to respond to the threats from these carriers by offering capacity controlled discount fares for the first time (i.e. seat management). This in turn led to a response by the chartered carriers to request the ability to offer frequent, if not scheduled, airline services. Push came to shove and by 1978 discount fares were available extensively throughout the United States and charter and new entrant carriers were being authorized to provide scheduled services.

Forces within Canada

Many of the same forces were present in the Canadian airline industry as well, although with a lag. Because of the lag, Canada experienced three additional forces which drove its deregulation. First, it could readily observe the impact that US airline deregulation was having. Canadian consumers watching US television stations and reading US print media, were immediately aware of the bargain airfares available in the US because of deregulation. They put pressure on politicians to implement similar reform in Canada, largely through the Consumers Association of Canada – see Reschenthaler and Roberts (1979). Second, the majority of Canadians live within driving distance of airports in the United States. Canadians were thus able to (and did) divert to US gateways for their air trips. This diversion took two forms: (1) Canadians travelling to an ultimate destination in the US (or elsewhere in the world) would drive to a US airport; and (2) when making a choice of destination for leisure trips, the availability of low airfares in the US induced Canadians to choose US over Canadian destinations. Third, some diversion of domestic Canadian traffic to trans-continental US routes was observed. For example, one tour operator in 1977 provided bus transportation from Toronto to an airport in the US, from which the consumer would fly to Seattle and then take a bus to Vancouver.

In hindsight, it seems unlikely that Canada would have deregulated its airline industry had the US not deregulated. It was primarily the demonstration effect of US deregulation on Canadian consumers and policy makers, as well as the diversion of air traffic from Canadian carriers that provided the forces which made Canadian airline deregulation necessary and inevitable.

5.4 The Canadian 'evolution' towards deregulation

It is difficult to date the beginning of the movement toward deregulation in Canada which officially began on 1 January 1988. Moreover, the process was not a smooth one as we shall document; there was some 'backsliding' on the 'road to Jerusalem'. We shall use 1979 as the 'starting point'.

On 23 March 1979, the federal Minister of Transport (with a federal election scheduled for 22 April) removed *all* capacity constraints on CP Air's share of the transcontinental market. The process of removing the constraints on CP Air began in 1959 when CP Air was *first* allowed to compete with Air Canada on transcontinental routes, but only with one flight per day each way between Vancouver and Montreal. Recall that Air Canada, by Cabinet order, had enjoyed a monopoly on transcontinental routes from 1937 to 1959. In August 1979, Wardair (Canada's largest all-charter carrier)[16] was licensed to provide *domestic*, advance booking charters (ABCs) in transcontinental markets. When it began service in May 1980 it competed with the scheduled airlines' charter class fares introduced two years earlier as part of their strategy of seat management. At the end of 1979, amendments were made to the Air Carrier Regulations by the Air Transport Committee of the Canadian Transport Commission which reduced the restrictions on domestic charter operations. These changes had been strongly 'pushed' by Wardair, controlled by colourful entrepreneur Max Ward. Then in February 1980[17] an Order in Council initiated by the Minister of Transport, Don Mazankowski, further reduced restrictions on domestic ABCs, e.g. it allowed one-third 'top off' sales during the last few days before the flight. The changes were widely seen as an effort to help Wardair compete with Air Canada and CP Air.

The juxtaposition between the old and new approaches to airline regulation could be seen in three documents produced in 1981. In February, the Department of Transport (1981a) published its study, *Economic Regulation and Competition in the Domestic Air Carrier Industry*, which proposed to continue with restrictive economic regulation that originated in 1938. In June, the Economic Council of Canada (1981) in its final report on the regulation reference of 1978 recommended phased deregulation of the airline industry. In August, the Department of Transport (1981b) released its report *Proposed Domestic Air Carriers Policy (Unit Toll Services)*. In summary terms, the DOT proposed to reinforce traditional price and entry controls and entrench the system which differentiated and 'co-ordinated' the roles of three 'tiers' of air carriers (two national, four regional and about 78 third-level or commuter carriers).

In April 1982, the House of Commons Standing Committee on Transport (1982) released its report on the DOT's (August 1981) proposals. The Committee 'rejected U.S.-style deregulation or even the more cautious five-year phased deregulation as proposed by the Economic Council (1981). Instead, the Committee endorsed the regulatory control of the CTC within a set of policy guidelines that would continue the "evolutionary process" by which greater, but controlled, competition had occurred' (Reschenthaler & Stanbury, 1983: 213).

In the face of a recession in the summer of 1982, Air Canada and CP Air offered deep discount (i.e. charter class) fares during a season in which fewer discount fares are usually available. Further, Air Canada reduced the advanced booking period to one day and removed the minimum and maximum stay requirements. Not surprisingly, the fraction of discount traffic rose, although it was well below the level in the US at the time. Both national carriers suffered substantial losses in the first half of 1982 (Reschenthaler & Stanbury, 1983: 215). On 19 August 1982 the CTC proposed new rules governing fares discounted more than 25% below the lowest unrestricted fare effective 1 November.[18]

The federal Minister of Transport, Lloyd Axworthy, formed an inter-departmental task force on airline regulation in December 1983. In September he had requested the CTC to hold hearings to review domestic charter fares. The Air Transport Committee of the CTC began hearings on regulation of domestic ABCs and inclusive tour charters in mid-February 1984, but they soon turned into hearings on the broader issue of deregulation. In March, the Minister of Transport instructed the Chairman of the Board of Air Canada to temper the senior executive's previously overt criticism of the Minister's proposed policy of liberalized regulation.[19] On 3 April, the executive vice-president of Air Canada, Pierre Jeanniot, accepted the principle of phased deregulation over six to 10 years. The result was that the officials of DOT were left supporting a policy of continued protectionism that was no longer desired by Air Canada!

On 10 May 1984 the Minister of Transport (Lloyd Axworthy) announced The New Canadian Air Policy.[20] It eased entry conditions and gave carriers more freedom to lower fares.[21] Air Canada was directed to divest its 85% shareholding of Nordair, the second largest of the four regional carriers. Under the new policy Canada was divided into two regions: north and south (with 95% of the population) (see Figure 5.1). All of the old regime rules applied to Northern Canada.

Unlimited entry was allowed into round-trip charter markets in Southern

Canada. Incumbent carriers in the South could exist freely when faced with new entry. In general, exit from the market continued to follow the PCN ('public convenience and necessity') criterion except when faced with new entry. All restrictions on conditions of service on airline routes were removed. This involved the removal of capacity, equipment type, number of stops, and frequency of service constraints on any given route. In southern Canada there was unlimited freedom to *reduce* prices. Price increases were allowed up to the rate of inflation of an input price index excluding labour (larger increases could be sought from the CTC). Mandatory booking and travel restrictions on discount fares were to be removed. Air Canada was restricted from engaging in anti-competitive pricing and scheduling practices unless private carriers were doing the same. Air Canada was told it could not receive funds from the federal government unless it met 'accepted financial tests'. The Regional Air Carrier Policy, described in section 5.2, was ended.

Over the next two years, the effects of the New Canadian Air Policy included the following: consolidation of licences; removal of restrictions on licences of former regionals; introduction of competitive business class services (e.g. Attache by CP Air in November 1984); sale of Nordair (owned by the federal government through Air Canada) to the private sector; Air Canada began to talk about partial privatisation; former regionals entered transborder routes; and introduction of frequent flyer programmes by the two national carriers in mid-1984. For more detail, see sections 5.5 and 5.7.

On 4 September 1984 the Progressive Conservatives under Brian Mulroney were elected with a huge majority. On 28 September the Air Transport Committee of the CTC (at the request of the Minister in May) established two zones of fare flexibility for class I, II and III carriers. Ellison (1985) described these changes as a shift from 'eroded protectionism to differentiated liberation'.

The new Conservative Minister of Transport, Don Mazankowski, issued the policy paper *Freedom to Move* on 15 July 1985.[22] Virtually complete deregulation of the airline industry was proposed. For example, entry was to be based on a 'fit, willing and able' test to be met by DOT safety requirements for operating certification, and adequate insurance coverage. There was to be no regulation of entry into routes. Only advance notice was to be required for exit from routes. There was to be no ongoing regulation of fares but appeal could be made to the new regulatory agency over *increases*. There was to be no control on the type or method of financing (lease versus ownership). This eliminated the ban on ownership of jets by third level carriers. The North–South distinction in the New Canadian Air Policy was to be eliminated.

The Tories were not able to enact legislation embodying total deregulation, nor were they able to legislate quickly. Bill C-126 the National Transportation Act, 1987 was given a first reading on 26 June 1986. It was reintroduced as Bill C-18 on 4 November.[23] On 27 April 1987 the House of Commons Standing Committee on Transport proposed a set of amendments, most of which were incorporated into the bill as enacted in the fall.

The National Transportation Act, 1987 came into effect on 1 January 1988.

Among the Act's general principles were the following:

- Safety is not to be compromised by changes in economic regulation. The highest practicable safety standards are to be maintained.
- The transportation system exists to serve the needs of shippers and travellers.
- Competition and market forces are to be the prime agents of providing an efficient and adequate transportation system.
- Economic regulation is to be minimized in order to encourage competition within and between modes.

Canadian air carriers are subject to two distinct regulatory regimes based on the location of the points they serve. In (and to) the North (officially, the 'designated area') carriers are subject to controls over entry, fares and other terms and conditions of service, although with reverse onus for the burden of proof. In the South,[24] which contains 95% of Canada's population, there is almost complete deregulation:

- Entry to the industry is based on the 'fit, willing and able' test comprised of objective safety and insurance requirements, and 75% Canadian ownership (and control in fact).
- Licence restrictions were abolished.
- Exit or reduction in frequency below once per week requires only 60 days' notice.
- Fare levels and decreases are not subject to regulation. Carriers must publish fares and cannot charge above the published level.
- Fare increases on monopoly routes are appealable to the new National Transportation Agency.

The regulatory regime that applies to Northern Canada (the 'designated area') is as follows:

- Entry to the industry is based on the 'fit, willing and able' test.
- In the case of new entry to a route, interveners have to show there would be a 'significant decrease or instability in the level of domestic service'.
- The Agency can limit types of service, aircraft size and type, routes, points, and schedules through restrictions on licences.
- Fare levels and increases are appealable to the Agency.
- Sixty days' notice is required for droping a route.
- Subsidies are possible for 'essential' existing services that cannot be provided on a purely commercial basis. These are to be awarded on the basis of competitive bids.
- The Agency can amend the definition of 'designated area' by altering its southern boundary, but also designate any area outside the designated area to be within it.

With respect to international commercial aviation, the new Act included the following provisions:

- International routes, including Canada–US transborder routes, will continue to be based on bilateral international agreements.
- The Minister is to seek fewer restrictions on the terms of bilateral agreements.
- The Minister is given a directive power regarding licensing to respond to actions prejudicial to Canada's interests.

Finally, the following general provisions are relevant to the airline industry:

- Carriers are to receive, as far as practicable, compensation for imposed public duties. Compensation is to be 'fair and reasonable'.
- The new National Transportation Agency which is made up of nine members must include at least one representative from each of five defined regions.
- The minimum size of reviewable mergers was set at $10 million in assets or gross sales in/from Canada.

5.5 Changes in industry structure

Consolidation of the trunk carriers

Table 5.1 shows that as deregulation was dawning in Canada in 1982 there were six carriers providing scheduled jet services. There are three things to note from the Table. First, the largest carrier, government-owned Air Canada, had over 50% of the total market by any measure. Second, the next largest carrier, CP Air, was less than half the size of Air Canada. Third, the remaining carriers were all very small. Three of the four were owned directly or indirectly by provincial or federal governments in Canada. What transpired over the next four years, was a consolidation which resulted in the merger of CP Air and the four small regional carriers and Wardair to form Canadian Airlines International Limited (CAIL) (see Table 5.2). It is useful to start this discussion with a review of the economic forces behind consolidation.

Forces leading to consolidation

Economies of scale: network size versus traffic density
White (1979) surveyed all major studies of the nature of airline costs and concluded that 'economies of scale are negligible or non-existent at the overall firm level'. Why, then, did the wave of airline mergers occur in both the US and Canada? The first reason is that a simple manufacturing industry concept of economies of scale is inadequate for modelling the relationship between inputs and outputs in this network-oriented service industry. Second, costs alone do not determine market structure. Demand is also relevant, and there are several aspects of demand that favour larger carriers.

Table 5.1 Description of the Canadian airlines carriers ranked by 1982 revenues.

	Ownership	1982 revenues (millions of Canadian $)	% of total	1982 passengers carried (millions)	% of total	1982 revenue tonne-kilometres (millions)	% of total	tonne-kilometres (millions)	% of total	1982 ratio of charter to total tonne-kilometres
Trancontinental carriers										
Air Canada	Federal government	2,171	56	11.77	52	2,559		53		5.5%
CP Air	Private (CP Ltd)	851	22	3.71	16	1,245		23		12.9%
Regional carriers										
Pacific Western Airlines (PWA)	Province of Alberta*	315	8	3.49	15	243		5		36.1%
Nordair	Air Canada	113	3	0.79	4	114		2		50.3%
Eastern Provincial Airlines (EPA)	Private†	93	2	0.79	4	75		2		4.1%**
Quebecair	Province of Quebec	69	2	0.61	3	48		1		43.3%
Charter carriers										
Wardair	Private	271	7	1.38	6	542		11		100.0%
Total		3,883	100%	22.54	100%	4,826	100%		100%	

* In December, 1983 all but 15% of PWA was sold to the public. A condition of the sale was that no person or group of persons other than the Province of Alberta control more than 4% of PWA's voting stock. Alberta has since reduced its holdings to 4%.
** EPA has reduced its charter traffic in recent years. This ration was 21%, 20% and 7% in 1978, 1979 and 1980 respectively.
† Ownership of EPA was transferred in 1984 from Newfoundland Capital Group to CP Air.
Sources: 1982 Traffic Statistics and 1982 Financial Statistics, International Civil Aviation Organization; Wardair *1982 Annual Report*; Eastern Provincial Airlines.

Table 5.2   Major changes in ownership of airlines in Canada during or following deregulation, 1983–90.

1983
- Province of Alberta sells 85% of its shares in Pacific Western Airlines (which it acquired in 1974) (December 1983).
- PWA acquired 42% of Time Air, the largest third-level (commuter) carrier (increased to 46.5% in 1989).

1984
- CP Air acquired Eastern Provincial Airways for $20 million.
- Innocan Inc acquired Nordair Inc from Air Canada which owned 85% of the shares since late 1978

1985
- Air Canada and Pacific Western Airlines each acquired 24.5% of Air Ontario.
- CPAL helps to establish Nordair Metro, a new feeder carrier, with a 35% interest.

1986
- CP Air acquired Nordair Inc.
- Nordair Metro acquired Quebecair (from the Province of Quebec).
- Air Canada acquired 49% of Air Nova Inc (July 1986).

1987
- PWA Corp acquired CP Air (for $300 million) to form Canadian Airlines International Ltd (CAIL) (announced in December 1986, but completed 30 January 1987).
- Air Canada acquired 75% of Air Ontario and 75% of Austin Airways through a holding company (announced in December 1986, completed in January 1987).
- Air Canada acquired Air BC* (February 1987, announced in December 1986).
- Time Air Inc. (46.3% owned by PWAC) acquired North Canada Air Ltd (March 1987).
- Air Canada and PWAC combine their computer reservation systems to form the Gemini Group, a 50:50 limited partnership (June 1987).
- PWAC acquires a 45% interest in Calm Air International Ltd (July 1987).
- PWAC helps establish Ontario Express Ltd with a 49.5% interest (July 1987).
- PWAC helps establish Inter-Canadian Inc (which combined Quebecair, Nordair Metro and Quebec Aviation) with a 31.4% interest (September 1987).
- PWAC increases CPAL's interest in Air Atlantic Ltd from 20% to 45% (September 1987).

1988
- Air Canada helps to create a commuter airline based in Quebec, Air Alliance, with 75% ownership (February 1988).
- Air Canada acquired 90% of Northwest Territorial Airlines (May 1988).
- Air Canada sells 43% of its shares to the public at $8.00 in a treasury issue (30.8 million shares) (September 1988).

1989
- Covia Partnership (half-owned by United Airlines) acquired a one-third interest in the Gemini Group with Gemini obtaining exclusive rights to Covia's computer reservations software (Apollo) in Canada (March 1989).
- CAIL acquires Wardair for $248 million (which was then a failing firm) (announced 19 January; completed 28 April).
- Balance of Air Canada shares (57%) sold in a wide distribution (July 1989) at $12.00.
- Inter Canadian (31.4% owned by PWAC) changes its name to Intair and seeks to sever its links with CAIL (October 1989).

1990
- Air Canada was reported to have outbid PWA Corp to acquire Air Toronto, a commuter carrier that also serves 10 US cities (*Financial Post*, 6 April 1990, p. 1). The sale had not been completed by the end of May 1990.

---

* Air Canada owned 100% to July 1988 and then 85% thereafter.

Caves *et al.* (1984) distinguish between airline economies of traffic density and economies of firm size. Under the latter output is expanded by adding points to the network; under the former output expands by increasing service within a given network (set of points served). Gillen *et al.* (1986) applied this concept to Canadian airlines, and developed it further by distinguishing between different types of airline traffic (scheduled, charter, freight). These and studies of other airlines reach a common set of conclusions.[25] Roughly constant returns to firm or network size exist for rather broad ranges of airline traffic. That is, adding or dropping cities from an airline's network does not raise or lower unit cost. In contrast, sizeable economies of traffic density seem to exist up to fairly large volumes of traffic. That is, adding more flights or more seats per flight on a given route will result in lower 'per seat' costs. However, once the minimum efficient traffic density level is reached, the curve is flat over a wide range, indicating that there are no more gains associated with greater density.

Demand side forces favour large carriers

Market equilibrium and therefore market structure is determined by the interaction of both supply (i.e. costs/production) and demand. In airline markets there are demand forces such that consumers prefer large airlines over small ones, all other factors such as prices being the same. In this context, large airlines mean those that serve a large number of points. Some of these forces have been present for some time, while others have been stimulated by marketing practices introduced since US deregulation.

In practice, there are at least three reasons why consumers prefer large airlines. One reason is due to information costs. A traveller knows that a large carrier can get him or her to just about anywhere in the country, while smaller carriers serve only a limited number of communities. Travel agents act as intermediaries for the consumer, but even here large network airlines have an edge, such as when an agent in one region needs to book flights in other regions.

A second reason consumers favour large airlines is attributable to the higher quality of service these airlines offer. If connections must be made, less of the traveller's time will be required with a single airline than when the trip involves switching airlines, because single airline flight connections are more likely to be timed to minimize waiting time at intermediate points (hubs).[26] Consumers are also aware that there is a lower probability of baggage being lost or delayed with a single airline, as well as a higher probability that the same airline's outbound flight would be held for a traveller on a delayed inbound flight.

The third factor causing consumers to favour larger over smaller carriers is the existence of frequent flyer programmes. These programmes reward the *individual* for patronizing a single carrier (even though the fare for business travellers may be paid by their employers). It is much easier to accumulate points with an airline that flies to a large number of destinations.[27]

Conclusion

In sum, there are natural market forces favouring large airlines in spite of evidence of constant returns to 'scale'. These are economies of traffic density, and in addition, the demand side factors such as information costs, higher quality travel, and reward programmes inducing consumers to favour large over small airlines. It appears that economies of traffic density can be fully exploited by an airline the size of CAIL. However, the traffic densities of the predecessors of CAIL were not sufficient to reach minimum efficient density.[28]

The formation of CAIL

In 1984, immediately following the announcement of the New Canadian Air Policy on 10 May 1984, CP Air acquired 100% of Eastern Provincial Airways, a regional serving the Maritimes[29] (see Table 5.3). Also, in May 1984 Air Canada was forced by the federal government to spin off its 85% interest in Nordair, which it did to a private non-airline entity. In 1986, Nordair indirectly acquired a minority interest in Quebecair from the Province of Quebec. CP Air acquired control of Nordair and thus CP Air had control of three of the four regional carriers by the fall of 1986. It attempted to acquire control of PWA, but was thwarted for a number of reasons, not the least of which was the Province of Alberta's requirement that no one individual could control more than 4% of the stock of PWA.[30] In the end, PWA acquired CP Air for $300 million and thus the consolidation of CP Air and the former regionals was complete (see Gillen

Table 5.3   Total aircraft seats in Canada, 1989.

	Aircraft	Total seats	% Total seats
Air Canada	115	17,697	37%
Air Canada feeder carriers (includes Air Toronto)	112	4,890	10%
Total Air Canada	227	22,587	47%
CAIL (including Wardair)	93	14,453	30%
CAIL feeder carriers (excludes Intair)	84	2,704	6%
Total CAIL	177	17,157	36%
Nationair (charter carrier)	10	2,635	6%
Worldways (charter carrier)	8	2,380	5%
Air 2000 (charter carrier)	1	180	0.4%
Intair	31	1,309	3%
City Express	8	360	1%
First Air*	19	911	2%
Total Canada**	481	47,519	

*   Data supplied by carrier. Excludes DHC Beaver fleet.
** Excludes Holidair, Vacationair and Odyssey (charter carriers) which became bankrupt in 1989/90. Does not include the many small and speciality operators.
Source: Tony Hine, *Transportation News*, December 1989 (Toronto: Scotia McLeod Securities).

et al., 1988).[31] In 1987 the name Canadian Airlines International Limited (CAIL) was chosen for the consolidated operation.[32]

In 1986 the former all-charter carrier, Wardair, was allowed to start scheduled airline services within domestic Canadian markets. It mounted a valiant effort to compete with Air Canada and CAIL, but by 1988 it was losing significant amounts of cash. In January 1989 Wardair was sold to PWA Corp for $250 million and its operations were merged into those of CAIL in the summer and fall of 1989.

The end result of this was that Canada had a duopoly in scheduled airline services.[33] Air Canada ended 1989 with somewhat over a 50% share of the market. CAIL had the corresponding slightly less than 50% market share.[34] Including international, cargo and other operations, Air Canada had 1989 revenues of $3.67 billion,[35] while PWA was only 72% of its competitor's size, with revenues of $2.6 billion.[36] In addition to controlling the trunk markets, the duopolists also controlled most of the feeder carriers in Canada as is discussed in the next section. There are a number of independent small turbo-prop operators as well as a handful of charter carriers still operating in Canada, but as seen in Table 5.3 the independents account for only 15% of the total seats available in Canada, and a much, much smaller percentage of total industry revenue.

Consolidation of the feeder carrier network

As happened in the United States, both Canadian trunk carriers developed feeder networks (see Table 5.2). In 1982 none of the scheduled jet carriers had a financial interest or feeder arrangement with any turbo-prop feeder carrier. By 1988, both Air Canada and CAIL had developed feeder networks which covered Canada from east to west and north to south. Table 5.4 describes the two feeder networks. Whereas CAIL's policy has been one of taking a large but minority equity stake in its feeders, Air Canada's policy has been to obtain majority ownership and therefore control of the feeder.

In October 1989, one of CAIL's feeders, InterCanadian, announced that it was terminating its feeder agreement with CAIL, and would act as an independent airline. Indeed, InterCanadian, now doing business as Intair, competes with CAIL, Air Canada and City Express[37] in the important Toronto–Montreal–Ottawa market. This defection has had several important consequences. First, it may have induced Air Canada to attempt to acquire majority ownership of its only minority-owned feeder, Air Nova.[38] Second, it creates a gap in CAIL's feeder network. InterCanadian was the second largest of CAIL's feeders. However, much of InterCanadian's former territory had been overflown by CAIL feeder Air Atlantic, thus this carrier may be able to help cover the slack. Third, there have been rumours that Intair may align itself with a US airline. If true, this could seriously threaten CAIL and Air Canada in important transborder markets. Fourth, a dispute arose in late 1989 regarding

Table 5.4   Feeder carriers in Canada, 1989.

	Region	Aircraft	Seats	% owned by trunk	Year acquired
Air Canada feeders					
Air BC	West	32	1,237	85	1986
Air Ontario	Ont/Man	43	1,951	75	1987
Air Alliance	Quebec	7	259	75	1988
Air Nova	Maritimes	14	749	49	1986
Northwest Territorial	W. Arctic	9	530	90	1986
First Air	E. Arctic	21	1,300	alliance only	
Air Toronto*	US	7	164	100	1990
Sub total		133	6,190		
CAIL feeders**					
Time Air	West	35	1,334	46.5	1983
Calm Air	Manitoba	14	276	45	1987
Ontario Express dba					
Canadian Partner	Ontario	20	554	49.5	1987
InterCanadian†	Quebec	31	1,309	35	1986
Air Atlantic	Maritimes	15	540	45	1985
Sub total		115	4,013		

*   Air Canada proposed purchasing Air Toronto in April 1990. The transaction has not been completed as this paper was written. Air Toronto's parent, Soundair, is bankrupt.

** CAIL provides its own services to the North via its Northern Services division.

†   InterCanadian terminated its feeder arrangement with Canadian in October 1989. Subsequently, CAIL sold its interest in the carrier, which does business using the name Intair.

ownership of slots at Toronto's Pearson International Airport. CAIL claimed ownership of the slots assigned at the previous slot allocation committee meeting. Intair required some additional slots in order to launch competing services. The dispute was eventually resolved by CAIL giving a few slots to Intair.

Intair's defection raises a critical issue about growth opportunities and financial prospects for feeder carriers. During the 1980s, these carriers enjoyed tremendous growth as they entered new markets, and took over services in communities being dropped by the trunk carriers operating jets. With deregulation entering a mature phase, and with the trunk carriers starting to take back some of their former markets (due to the stimulation of the market by the feeder carriers providing frequent service), the feeder carriers face an uncertain future in the 1990s. They will be prevented from competing directly with the trunks, and are hemmed in by other allied feeder carriers in adjacent geographic regions. Intair decided that its future as a feeder carrier for CAIL was limited, and it decided to strike out on its own.

A final comment should be made about the consolidation of the feeder airlines. Feeder airline markets in Canada are unable to support more than two carriers. Since almost every viable air market in Canada is now served by feeder carriers affiliated with the two trunk duopolists, a significant entry barrier exists. A potential entrant into the trunk carrier industry would need access to feeder traffic, but would find that this is 'owned' by the duopolists. This point was driven home by the experience of Wardair just prior to its sale in January 1989. Access to this feeder traffic was so important for the profitability of Wardair's trunk routes, that it decided to pay the feeder airline fares for individuals willing to connect to Wardair trunk flights. This was necessary for its survival on the trunk routes, but it involved a transfer of resources from Wardair to its competitors.[39]

Consolidation of the distribution/marketing channels

Even though airlines have the ability to sell tickets directly to consumers, 70% of airline tickets in Canada are sold by travel agents. While the agents are supposedly independent, they are paid by and agents of the airlines, not the consumers. Airlines have two methods of influencing agents' choices. The first of these is by paying higher than normal commission on airline tickets sold by the agents. This gives the agent an incentive to steer the consumer to the services offered by the airline paying the 'override'. Since these overrides are hidden to the consumer and sometimes to competing airlines, the effect can be very anti-competitive.

The second avenue airlines have for controlling decisions of agents is via computer reservation systems (CRSs); 80% of travel agents in Canada now use a CRS.[40] Studies in the United States have found that the way the information is displayed in a computer reservation system has enormous influence on consumer choices. American Airlines, for example testified to Congress that 92% of all ticket sales come from the first US screen displaying information on a given market. Further, 54% of sales come from the first line on the first screen. This creates an overwhelming incentive for the carriers to bias CRS displays of flight information to favour the flights of the airline owner of the CRS. Even if CRS displays are unbiased, a 'halo' exists which results in the agents favouring booking on the airline which owns the CRS.[41]

In the 1960s Air Canada developed the world's first airline computer reservation system. This system eventually came to be known as Reservec. Until 1984, it was the only CRS system available in Canada. In January 1984, CP Air, recognizing the problem that it was facing with its primary competitor controlling the travel agent portion of the distribution channel, launched a competing CRS system, Pegasus. However, CP Air quickly discovered that penetrating the market would be difficult at best. Air Canada had already locked up the major travel agents with its Reservec system. CP Air found that it could successfully market its Pegasus system only to the smaller agents. Further, while CP Air paid Air Canada a fee for every CP Air ticket sold on Reservec, Air Canada refused to make any payments to CP Air when an Air Canada flight would be booked on a Pegasus system. CP Air claimed that its Pegasus effort was failing, and approached Air Canada about merging the systems. Apparently Air Canada refused. CP Air then opened negotiations with American Airlines to bring its Sabre system into Canada as a replacement for Pegasus. Sabre dominated the US and was making significant penetrations elsewhere in the world. This threat appears to have been sufficient to get Air Canada to come to the bargaining table. Effective 1 June 1987 Air Canada and CAIL (the successor of CP Air) agreed to merge their two CRSs into a single system, Gemini. Gemini was then owned 50/50 by the two airlines. Gemini decided to abandon both carriers' home grown systems, and to replace them with a US system. An initial agreement was arrived at with TWA/Northwest's PARS, but this was eventually replaced with Gemini adopting United Airlines' Apollo/Covia technology. Covia became a one-third owner of Gemini.

The Gemini merger resulted in a consolidation of the CRS market in Canada. Gemini's Canadian market share is 90%, compared with 10% share for Sabre. Although the merger was attacked by the Bureau of Competition Policy under the Competition Act,[42] the case was settled with a Consent Order under which CAIL and Air Canada are required to provide complete, timely and accurate information on the data in its CRS to all other CRSs operating in Canada on the same basis as it is given to Gemini. Air Canada and CAIL were ordered to participate in all CRSs operating in Canada on commercially reasonable terms. They were ordered to make available to other CRSs in Canada the same advance seat selection and boarding pass capability which had been provided to Gemini. Further, Air Canada and CAIL were ordered to provide a 'look but not book' link (effective 31 January 1990) and a 'look and book' link to other CRSs (effective 30 June 1991). In addition, the Consent Order specified a set of rules for the operation of CRSs.

Non-scheduled air carrier developments

Prior to 1984 all of the scheduled airlines in Canada engaged in substantial charter services. In addition there were a handful of charter airlines, with Wardair dominating these. Pursuant to the New Canadian Air Policy of May 1984, Wardair began to operate scheduled services within Canada in April 1986, but only on a limited basis (six cities). This led to a perception that it was abandoning the charter market. Table 5.5 indicates that this was not true. From 1986 to 1988, Wardair actually increased the absolute amount of charter services it offered, although charters as a percentage of total services were declining. This is because Wardair accommodated scheduled services by expanding its fleet. In 1989, following its acquisition by CAIL, the absolute level of charter services provided by Wardair dropped by 25% but was still substantial. Nevertheless, a perception was created in the industry that Wardair was leaving the charter market in favour of the scheduled market and that this would create opportunities for new charter airlines.

A number of new charter airlines were launched between 1986 and 1988. In addition two previously established chartered carriers, Worldways and Nationair continued to provide services in the market. Combined, Nationair and Worldways have 5,035 seats compared with Wardair's 3,240. While

Table 5.5   Wardair charter services, 1983–89.

Year	Charter services	% of total revenue passenger miles
1986	1,933	47
1987	2,110	48
1988	2,509	44
1989 (9 months extended to annual)	1,871	35

Source: Company documents.

Worldways and Nationair operated with older aircraft, some of the new chartered carriers, such as Odyssey, operated with brand new leased aircraft. With new aircraft, these carriers were forced to meet high monthly aircraft ownership costs. If charter prices softened because of excess capacity in the market, these carriers would nevertheless be forced to continue flying in order to make some contribution to their ownership costs. Carriers with older aircraft may be able to weather the storm and park aircraft when operations are not profitable.

To further add to the charter capacity problem, Air Canada increased its presence in the charter markets as well. For example, in 1990 it increased the number of seats from Montreal/Toronto to Florida by 30%.[43] The consequence of the entry of the new charter carriers (especially those with new aircraft), Air Canada's increase of charter service, and Wardair's continued presence in the market, was the shaking out of five of the fledgling charter carriers in late 1989 to 1990.[44] Whether the amount of capacity shaken out of the market was sufficient remains to be seen.

It should be mentioned in passing that charter services continue to be important in Canada. In an earlier study, Gillen *et al.* (1985) claimed that 'many charter markets are an artificial product of regulation. They disappear in a deregulated environment'.[45] The US witnessed the virtual disappearance of its *domestic* charter market, as did Canada. However, air travel *between* Canada and the United States continues to be regulated, along with all other international services. Thus, even though there may be sufficient traffic to warrant scheduled services between Toronto and Tampa, Montreal and Fort Lauderdale, etc, charter services are the only way at present to serve the market. Should the US–Canada bilateral air agreement be liberalized, then charter services would be diminished. They would not disappear, as there would still continue to be an important charter segment between Canada and non-US sunspot destinations.

Creating brand loyalty

As deregulation began in the United States, air transport could largely be viewed as a commodity. That is, consumers had little loyalty to particular producers. Some airlines, such as People Express, followed the appropriate strategy for a commodity: follow a cost leadership strategy and compete on the basis of price. A few other airlines, notably American Airlines, decided to pursue strategies which would build brand loyalty and undermine the commodity nature of the service. The most notable of these strategies was the introduction of frequent flyer programmes by American Airlines in 1981. Economists would describe this process as one of putting some slope in the carrier's demand curve.

Because large carriers can offer frequent flyer rewards at lower costs, these programmes create a significant barrier to entry.[46] Frequent flyer programmes came to Canada in July 1984 only a few months after they were permitted under the New Canadian Air Policy. Prior to this they were not allowed by the government. The Canadian carriers introduced these largely in order to maintain market share on trans-border routes to the US as they were losing customers to the US carriers offering these reward systems.

In Canada, as in the United States, a trunk carrier awards points for travel on its affiliated feeder carriers. However, it never allows a competing carrier to join its frequent flyer plan. Non-aligned smaller carriers are also generally excluded from these plans. Because of their attractiveness to consumers, membership by an air carrier in the frequent flyer plan of a large carrier is almost required these days in North America. Both PSA in the US and Wardair in Canada cited frequent flyer programmes as a problem and as a major reason for their mergers into larger airline systems.

Summary on existing entry barriers

The Canadian airline industry has undergone dramatic consolidation. This includes consolidation at the level of the trunk carriers, at the level of the feeder carriers, and of the computer reservation systems. Potential entrants to the industry at this point will discover that there are significant, perhaps insurmountable, barriers to entry into the domestic Canadian airline industry.

First, only carriers the size of Air Canada and CAIL appear to be able to fully exploit the available economies of traffic density. Canadian markets are simply not large enough to support a third carrier of minimum efficient size. Second, the entrant would discover that the primary distribution channel is controlled by the two duopolist airlines, although their Gemini CRS is subject to a court order requiring it not to discriminate against any carrier or other CRS. Third, it would find that almost all feeder air carriers are aligned with one of the two duopolists. Thus a potential trunk entrant would not have access to the important feeder traffic from the smaller communities. Fourth, the existence of frequent flyer programmes and the demand side advantages associated with these, along with the inertia of consumers having been members of the duopolists' programmes for six years, creates a brand loyalty problem which it might not be able to overcome. Fifth, although this has not been discussed, the carrier would discover that at most airports in Canada there is a shortage of ticketing and to a lesser extent gate space. At two of the key Canadian airports, Toronto and Vancouver, there is a shortage of airside capacity. Toronto, like many airports in the US, has slot controls in effect. Finally, it should be pointed out that the foreign ownership limitation of the Canadian National Transportation Act prevents a US or other foreign airline (or non-airline) from bankrolling a potential entrant into the Canadian domestic market beyond 25% of the voting shares.

Overall, this is a formidable list of entry barriers. While any individual barrier perhaps could be overcome by one technique or another, the cumulative impact is likely to be such as to rule out any entry into the domestic Canadian airline industry for the foreseeable future.

5.6 Performance in the early deregulation era

In contrast to the United States, the Canadian airline industry underwent a gradual regulatory liberalization prior to the formal deregulation of the southern Canadian market which took effect on 1 January 1988. The 1978 US deregulation had an indirect influence on consumers' expectations and the behaviour of regulators and airline managers in Canada. Because of US deregulation, the Canadian Transport Commission (Air Transport Committee) was forced to respond, albeit very slowly, to the changing competitive conditions and rising consumer pressure for cheaper fares by approving some discount fares (mainly in the form of seasonal 'seat sales') and relaxing the conditions on charter fares, while keeping entry/exit and service conditions tightly regulated. In 1979, the Minister of Transport removed all restrictions on CP Air's capacity share of the transcontinental routes. Canadian airline managers also had the benefit of observing the consequences of US deregulation, and thus had time to adjust their operations, at least marginally, for the forthcoming competitive era.

The New Canadian Air Policy announced on 10 May 1984 (which eased entry conditions, gave carriers more freedom to lower fares, removed all restrictions on conditions of service in routes licences, and repealed the regional air carrier policy),[47] signalled the industry unequivocally that deregulation was bound to occur in the near future. Since then, carriers have responded vigorously by reducing and streamlining their work forces, taking tougher stands in negotiations with unions, adopting innovative marketing tools such as frequent flyer programmes and sophisticated pricing through seat management, rationalizing their network and aircraft fleet, thinking strategically to create national feeder networks, and strengthening their computer reservation systems.

Due to the gradual nature of regulatory liberalization and deregulation in Canada, this section examines the effect of deregulation on the industry's performance by comparing the following time periods as permitted by the data.[48]

1978 to present: US Deregulation
1984 to 1988:   New Canadian Air Policy
1987/88 to present: Deregulation in Southern Canada

Effects of deregulation on fares and service quality

Economists, notably Douglas and Miller (1974), argue that price regulation can lead to excessive quality competition in the airline industry. This in turn leads to higher fares, larger aircraft, higher frequency of service, and lower load factors than the levels socially optimal price–quality tradeoffs call for. Deregulation is expected to reduce these inefficiencies, reduce prices, lower frequencies of service, reduce aircraft size, and increase load factor. Below we discuss the changes in these statistics since the regulatory relaxation in 1984 and deregulation in 1988. It should be kept in mind, however, that Canada regulated conditions and quality of service, unlike the US. Thus, the Douglas and Miller predictions may be less relevant for Canada or other countries which regulated service quality.

Air fares
Table 5.6 shows that during the 1978–88 period, the average yield per passenger-kilometre increased by 68% in Canada while the increase was only 43% in the US. Part of the difference, however, is due to the higher inflation rate in Canada. In constant dollar terms, the average yields in Canada decreased by 18% as compared to a 22% reduction in the US. A breakdown into two sub-periods reveals that, in constant dollar terms, the US average yield decreased twice (8%) as fast as that of Canada (4%) between 1978 and 1984, while between 1984 and 1988 it decreased 15% versus a decrease of 10% in Canada. During the last two years (1988–89), the constant dollar average yield in Canada decreased by about 11 percentage points, while it increased in the US (at least for 1988).

Various yield statistics indicate that airfares in Canada have responded to increased competition brought about by the regulatory relaxation in 1984 and deregulation in 1988. Lack of Canadian data for estimating cost and demand functions renders it impossible to rigorously decompose the fare reductions into sources such as regulatory changes, technological advancement, and input price changes.

The utilization of discount fares increased from 15% in 1980 to over 60% in 1988 while comparable US figures are 57% in 1980 and 91% in 1988.[49] Figure 5.2 gives Canadian figures. Furthermore, the level of discount below the regular unrestricted economy fare increased gradually over time: from 25% in 1983 to 45% in 1988 (Figure 5.3). In 1989, it ranged from 40% to 50% in the domestic market, about 35% in the transborder market, 45% to 50% for sunspot destinations, and about 60% in other international markets. This compares to the US average of 63% discount off the coach fare in 1988 (Figure 5.4).

Table 5.6 Index of average yields (Canada–United States).

Year	Average yield per pass-km (C cents)	Canada Index current $	CPI	Fare index constant $	Average yield per pass-mile (US cents)	United States Index current $	CPI	Fare Index constant $
1977					8.4	101	93	109
1978	5.7	100	100	100	8.3	100	100	100
1979	6.0	105	109	96	8.7	105	111	94
1980	6.8	119	120	99	11.0	133	126	106
1981	8.1	142	135	105	12.3	149	139	107
1982	8.6	151	150	103	11.8	143	148	96
1983	9.1	160	158	100	11.6	140	153	91
1984	9.1	160	165	96	12.1	147	159	92
1985	9.4	165	172	96	11.7	141	165	85
1986	9.5	167	179	93	10.6	128	168	75
1987	10.0	175	187	93	11.0	133	174	76
1988	9.6	168	194	86	11.8	143	181	78
1989	9.6	168	204	82				
% changes								
1978–84		60%	65%	– 4%		47%	59%	– 8%
1984–88		5%	18%	– 10%		– 3%	14%	– 15%

Compiled from:
Average Yields: Air Transport Association of America, *The Annual Report of the US Scheduled Airline Industry*, (Washington, DC, various issues).
Statistics Canada, Catalogue Nos, 51–006 & 51–206, *Air Carrier Operations in Canada*, various issues.
CPI: IMF, *International Financial Statistics*, yearbook, 1989.

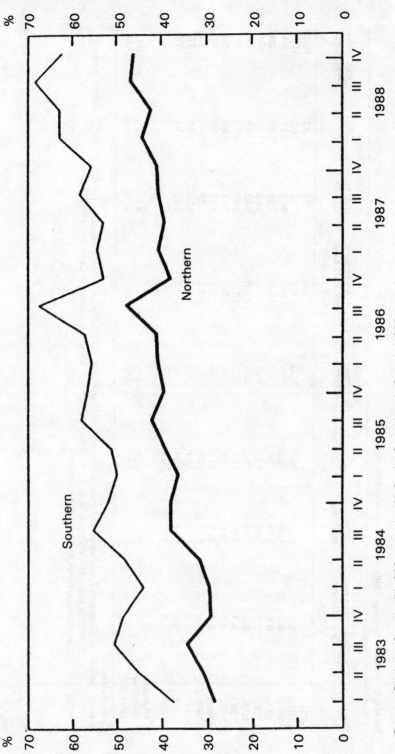

Source: Statistics Canada, 51–002 quarterly, *Air Carrier Operations in Canada* (1988 issues).

Figure 5.2   Discount fare utilization, by sector, 1983–88.

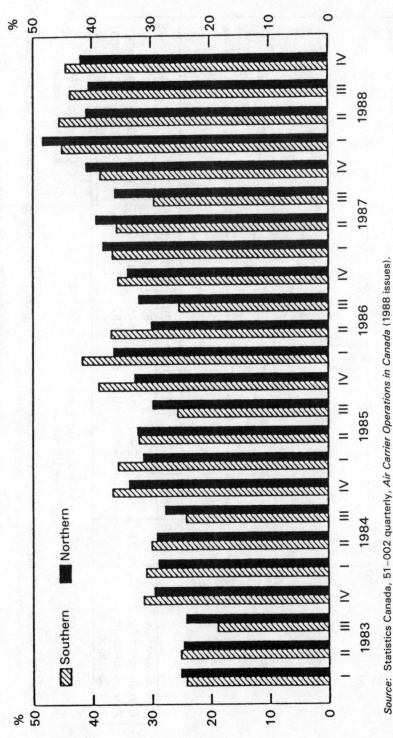

Source: Statistics Canada, 51–002 quarterly, *Air Carrier Operations in Canada* (1988 issues).

Figure 5.3 Discount off the economy fare, by sector, 1983–88.

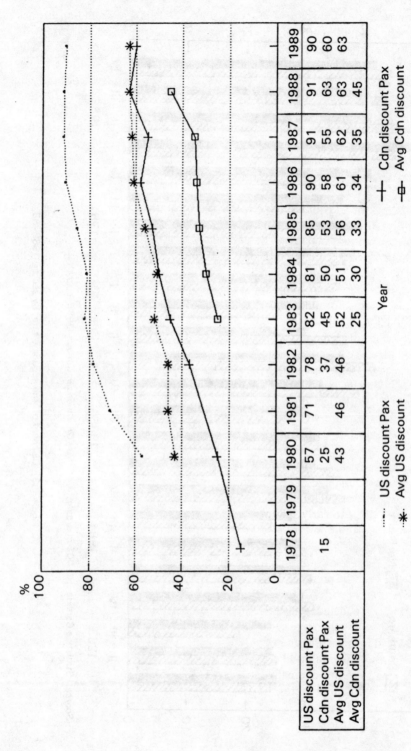

	1978	1979	1980	1981	1982	1983	1984	1985	1986	1987	1988	1989
US discount Pax	15		57	71	78	82	81	85	90	91	91	90
Cdn discount Pax			25		37	45	50	53	58	55	63	60
Avg US discount			43	46	46	52	51	56	61	62	63	63
Avg Cdn discount						25	30	33	34	35	45	63

Year

---- US discount Pax   —+— Cdn discount Pax
····* Avg US discount   —□— Avg Cdn discount

Figure 5.4   Domestic discount fares and utilization (Canada–US comparison).

Source: Compiled from ATA publications, *Airlines 1989 – A Year-end Review*. Statistics Canada 51–002, 51–004 and figures supplied by Lisa Di Pietro, Senior Statistician, Aviation Statistics Centre, Statistics Canada.

An examination of the data by market segment reveals that Canadian airlines have responded differently in different markets depending on the extent of competition. The economy fare for the domestic market increased between 1985 and 1989 while decreasing substantially for transborder, sunspot destinations and other international markets. Although there are substantial seasonal variations in discount fare levels, the discount fares for sunspot and international routes have decreased significantly faster than the domestic and transborder routes. For example, the third quarter statistics indicate that between 1985 and 1989, discount fares for domestic and transborder markets decreased only by 4% and 6%, respectively, as compared to the reductions by 14% and 13%, respectively, for sunspot destinations and other international routes.

The northern Canadian market has significantly lower utilization (or availability) of discount fares (slightly over 40% in 1988 as compared to about 60% in the southern market). The average percentage of discount below the regular unrestricted economy fare is only marginally lower in northern Canada than in southern Canada. In addition, airfares for short-haul routes have increased far more than long-haul routes. This was to be expected, as the regulatory pricing formula kept per mile prices in short-haul markets below costs (while long-haul fares were held above costs). The rise in short-haul fares may also reflect discriminatory pricing practices by airlines taking advantage of differential price elasticities between short and long distance travel.

As expected, Canadian air carriers responded differently to different market segments taking into account the differential price elasticities in these segments, as well as the historical distortion of regulated rates relative to costs. Price discounting was more vigorous in international markets, long-haul domestic routes, and sunspot destinations where competition is more intense than in other markets. Airlines also adopted dynamic pricing techniques through sophisticated seat inventory management systems, in order to maximize total revenue for a given flight. This became possible due to their new found pricing freedom under deregulation.

From the 1970s to early 1989 when it was acquired by PWA Corp (CAIL), Wardair played an important role in increased fare competition, especially for long-haul markets. The merger of CAIL and Wardair in 1989 generated widespread concern over anti-competitive consequences of the duopoly system. Some argue that the presence of two strong competitors may be better than the pre-1984 situation with Air Canada and many weaker carriers.[50] Morrison and Winston (1989) concluded, using US post-deregulation airfare data, that fares would increase less than one cent per mile if the number of actual competitors were reduced by one from an initial level greater than two. However, the effect of elimination of Wardair is likely to be higher than they would predict, because Canada does not have credible potential competitors in transcontinental markets like the United States.

Service quality

There are many dimensions of quality in airline services. These include frequency of service, size of aircraft, inflight services, schedule reliability, safety, direct vs. stopover flight, interline vs. intraline transfer, etc. These are now discussed.

Departure frequency: Morrison and Winston (1986: 24–36) found that increased flight frequency (and the associated reduction in schedule delay time) is the most important source of welfare gain from US deregulation. Table 5.7 reports average weekly departure frequencies of scheduled services on domestic Canadian routes. Figures are reported separately for the north and the south, and by jet and non-jet aircraft. Average weekly departures almost doubled in southern Canada, and more than doubled in the north during the 1984–89 period. It had remained virtually unchanged between 1978 and 1984. It is noted that in the north, jet aircraft frequency decreased by 25% between 1978 and 1989 while non-jet aircraft departures increased by 246% during the same period. However, even in the north the frequency of jet aircraft departures increased by 10% between 1984 and 1989.

Deregulation of the south and the regulatory relaxation in northern Canada appear to have increased schedule frequencies in both markets. Since 1984 the total schedule frequency in the north and the south increased by 114% and 93%, respectively.

Airline safety: Opponents of airline deregulation argued that deregulation could jeopardize safety as firms cut costs at the expense of safety. Others, including Morrison and Winston (1989), argued that market forces provide strong incentives for carriers to conduct safe operations. Statistics reported by Morrison and Winston (1989) show that although the accident rate ('fatal accidents per 100,000 departures') has year-to-year variations, there is a clear declining trend in the accident rate since commercial aviation began, and this trend has continued after the 1978 deregulation. Andriulaitis *et al.* (1986: 48–52) also concluded that the number of fatal accidents both per 100,000 aircraft hours and per 100,000 departures has decreased since US deregulation. Figure 5.5 reports that the number of accidents per 100,000 hours flown for the Canadian airline industry decreased from 13.6 in 1981 to 9.4 in 1989. Furthermore, safety of the transportation system as a whole has improved far more than the reduced airline accident rate would indicate, because airline deregulation diverted many travellers out of their more dangerous automobiles and into aircraft. Concerns about air safety have also risen during the past few years because of the increasing near-miss cases that are being reported, shortages of air traffic controllers (in Canada and the US), and crowding of airways and airfields. Although these concerns must be addressed soon, these problems are not caused by deregulation.

Other dimensions of service quality: Clearly, the regulatory changes have

Table 5.7 Aircraft departure frequencies (commercial operations).

Departures per week				Domestic South			
	1978	1983	1984	1985	1986	1988	1989
Jet	4422	4539	4485	4788	4922	5412	5451
Non-jet	2490	2363	2890	3611	5294	7926	8843
Total	6912	6902	7375	8399	10216	13337	14293
Departures per week (normalized)							
	1978	1983	1984	1985	1986	1988	1989
Jet	100	103	101	108	111	122	123
Non-jet	100	95	116	145	213	318	355
All aircraft	100	100	107	122	148	193	207

Departures per week				Domestic North			
	1978	1983	1984	1985	1986	1988	1989
Jet	728	586	497	489	501	480	543
Non-jet	774	919	998	1149	1610	2475	2676
Total	1501	1505	1494	1638	2111	2956	3219
Departures per week (normalized)							
	1978	1983	1984	1985	1986	1988	1989
Jet	100	81	68	67	69	66	75
Non-jet	100	119	129	149	208	320	346
All aircraft	100	100	100	109	141	197	214

Departures per week				Domestic Total			
	1978	1983	1984	1985	1986	1988	1989
Jet	5150	5125	4982	5277	5423	5892	5994
Non-jet	3263	3282	3887	4761	6904	10401	11518
Total	8413	8407	8869	10037	12327	16293	17512
Departures per week (normalized)							
	1978	1983	1984	1985	1986	1988	1989
Jet	100	100	97	102	105	114	116
Non-jet	100	101	119	146	212	319	353
All aircraft	100	100	105	119	147	194	208

Source: *Air Transport Monitor* vol. 1 Nos 1–4, 1985; vol. 2 Nos. 1–4, 1986. *Aviation Industry Review*, first–fourth quarter 1989.

influenced carrier fleet decisions. As in the United States, Canadian carriers are now using smaller and more economical aircraft than in the past as they try to increase schedule frequency in an efficient manner. The two major carriers transferred low density routes to their respective feeder carriers as the latter can serve those markets more efficiently with small aircraft. Hub-and-spoke operations have reduced the percentage of unbroken trips. Connections at hub airports inconvenience travellers. However, as both Air Canada and CAIL have set up feeder networks throughout Canada, most passengers can avoid inconvenient inter-line transfers. Increased congestions delays at hub airports in Toronto and Vancouver have reduced the schedule reliability of airline

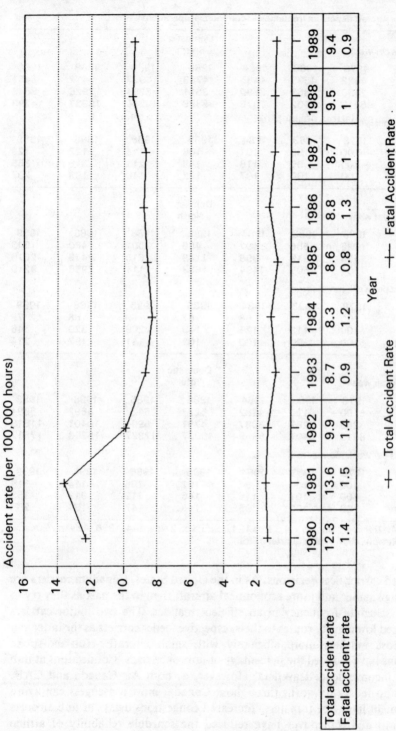

	1980	1981	1982	1983	1984	1985	1986	1987	1988	1989
Total accident rate	12.3	13.6	9.9	8.7	8.3	8.6	8.8	8.7	9.5	9.4
Fatal accident rate	1.4	1.5	1.4	0.9	1.2	0.8	1.3	1	1	0.9

Source: Canadian Aviation Safety Board *Annual Report*, 1989, Table 2.

Figure 5.5   Air safety: Canadian commercial operations.

services. However, the congestion delay problems in Canada are less severe than in the United States, partly because the geographic and demographic nature of the country limits the extent of efficient hub-and-spoke operations in Canada.

Effects of deregulation on employment, profits and welfare

Employment and wages
Figure 5.6 plots the number of employees for the level I carriers and the industry (levels I–III carriers). Employment decreased substantially between 1981 and 1984 (from 142,000 to 127,000 for the industry, and from 122,000 to 109,000 for the level I carriers), while it increased substantially in the 1984–88 period. The reduction in the 1981–84 period was largely caused by the economic recession, although it may have been aggravated by efficiency-conscious airline management influenced by US airline deregulation. Airline employment increased since 1984 largely due to two factors: improvement of the economy, and traffic stimulated by lower fares. Employment increased more rapidly in the levels II and III carriers than level I carriers. This result is similar to the situation observed after the 1978 US deregulation (see Andriulaitis *et al.*, 1986). Real average wages per employee remained virtually unchanged during the 1978–88 period in the range of 19,000 to 21,000 in 1978 dollars, while nominal wages have grown each year (Figure 5.7).

Carrier profits
Data on operating and net income for the level I carriers and the levels I–III carriers in Canada indicate that the airlines in Canada lost money during the severe 1982–83 recession. Similarly, the US scheduled airlines suffered operating losses of $222 million, $455 million and $733 million in 1980, 1981 and 1982, respectively (see Andriulaitis *et al.*, 1986: 125). Canadian airline profitability has improved considerably since 1984. Figure 5.8 reports the ratio of operating profit to operating revenue and the ratio of net profit to operating revenue for the level I carrier industry and for levels I–III. The operating profit ratio (markup), and the net profit ratio were higher prior to the regulatory relaxation in 1984 than after although total operating profits were higher in recent years than in the late 1970s. After deregulation, airline profits are spread over larger volumes of traffic because of rising competition.

Productivity
Gillen *et al.* (1985) measured and analyzed Canadian airline productivity up to 1981. Productivity measurement and comparison is a very complex and data-intensive undertaking beyond the scope of this study. Nevertheless, we are able to provide some simple productivity measures here. In order to compute factor productivity, output and input quantities must be determined. Total revenue-tonne-kilometres (RTK) is used here as a very crude measure of output. Number

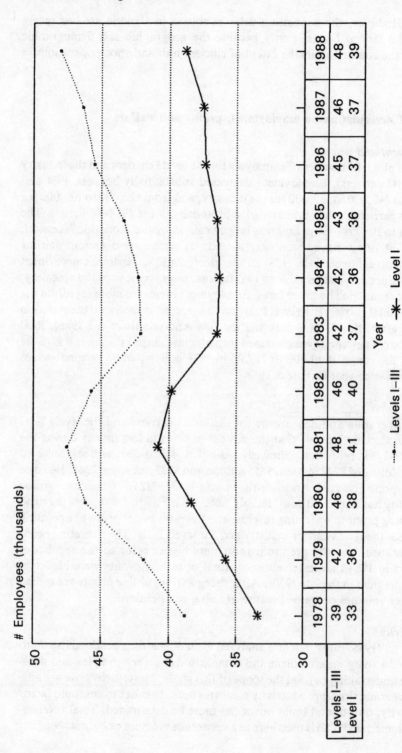

Year	1978	1979	1980	1981	1982	1983	1984	1985	1986	1987	1988
Levels I–III	39	42	46	48	46	42	42	43	45	46	48
Level I	33	36	38	41	40	37	36	36	37	37	39

Source: Compiled from Statistics Canada 51–002, *Air Carrier Operations in Canada.*

Figure 5.6   Number of employees: Canada level I and levels I–III carriers.

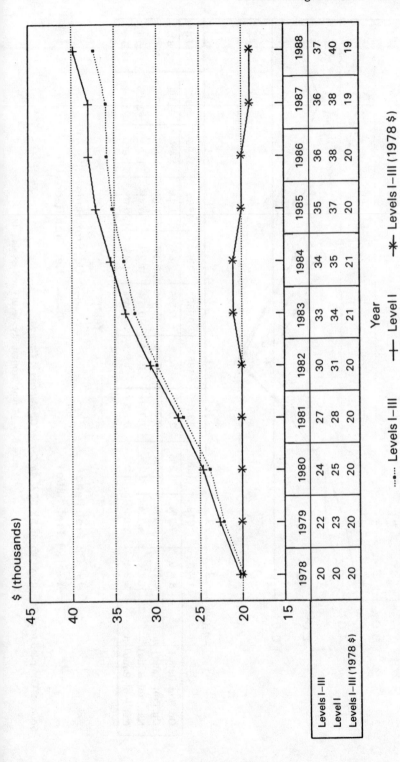

	1978	1979	1980	1981	1982	1983	1984	1985	1986	1987	1988
Levels I–III	20	22	24	27	30	33	34	35	36	36	37
Level I	20	23	25	28	31	34	35	37	38	38	40
Levels I–III (1978 $)	20	20	20	20	20	21	21	20	20	19	19

Figure 5.7 Average annual wage per employee: level I and levels I–III carriers.

Source: Compiled from Statistics Canada 51–002, *Air Carrier Operations in Canada.*

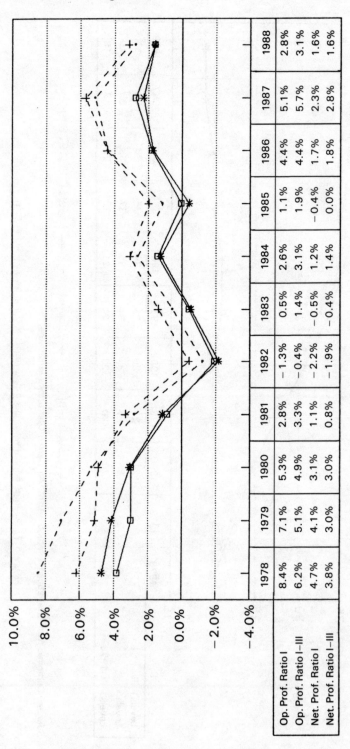

	1978	1979	1980	1981	1982	1983	1984	1985	1986	1987	1988
Op. Prof. Ratio I	8.4%	7.1%	5.3%	2.8%	−1.3%	0.5%	2.6%	1.1%	4.4%	5.1%	2.8%
Op. Prof. Ratio I–III	6.2%	5.1%	4.9%	3.3%	−0.4%	1.4%	3.1%	1.9%	4.4%	5.7%	3.1%
Net. Prof. Ratio I	4.7%	4.1%	3.1%	1.1%	−2.2%	−0.5%	1.2%	−0.4%	1.7%	2.3%	1.6%
Net. Prof. Ratio I–III	3.8%	3.0%	3.0%	0.8%	−1.9%	−0.4%	1.4%	0.0%	1.8%	2.8%	1.6%

Year

‑‑‑ Op. Prof. Ratio I   + Op. Prof. Ratio I–III

—✳— Net. Prof. Ratio I   ‑☐‑ Net. Prof. Ratio I–III

Source: Compiled from Statistics Canada 51‑002, *Air Carrier Operations in Canada.*

Figure 5.8   Airline profitability: level I and levels I–III.

RTK per employee (thousands)

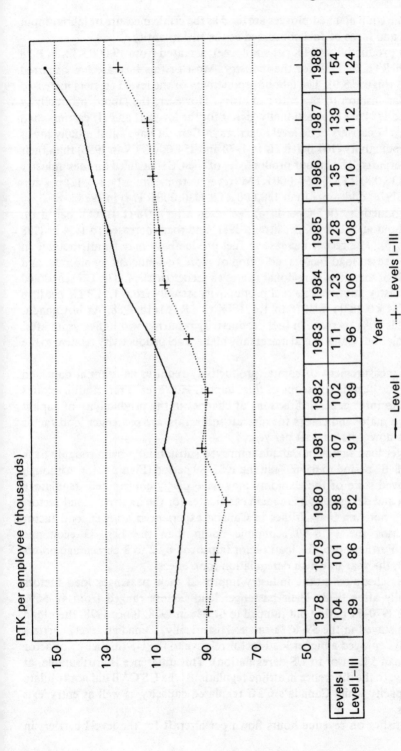

	1978	1979	1980	1981	1982	1983	1984	1985	1986	1987	1988
Level I	104	101	98	107	102	111	123	128	135	139	154
Level I–III	89	86	82	91	89	96	106	108	110	112	124

Year

—•— Level I   ⋯+⋯ Levels I–III

Levels I and II figures include levels I and II carriers for consistency of classification.

1978–80 figures include levels I–III carriers.

Source: Compiled from Statistics Canada 51–002, *Air Carrier Operations in Canada.*

Figure 5.9 Labour productivity: level I and levels I–III carriers.

of full-time equivalent employees are used as the crude measure of labour input quantity, and litres of fuel consumed as the fuel quantity.

Labour productivity (RTK per employee) increased from 89,000 RTK in 1978 to 124,000 RTK in 1988 for the industry. Most of this increase has occurred since 1983 (Figure 5.9). The labour productivity of the level I carriers increased in a similar fashion to the entire industry. However, the labour productivity gains since 1984 are proportionally higher for the levels II and III carriers than for the level I carriers. The level I carriers in Canada have substantially lower labour productivity (104,000 RTK in 1978 and 154,000 RTK in 1988) than their US counterparts. The labour productivity of the US scheduled airlines industry is about 136,000 RTK ($= 94,000$ RTM (revenue-tonne-miles) $\times 1.4512$) per employee in 1978, and more than 188,600 RTK (130,000 RTM) in 1988.

Fuel productivity (RTK per litre) decreased after 1978 (1.41 RTK per litre), bottomed out at 1.22 RTK per litre in 1981, and then increased to 1.54 in 1988 (Figure 5.10). The recent increase in fuel productivity may be attributable in part to increased load factors, adoption of more fuel-efficient jet aircraft, and the spinoff of low density regional routes to feeder carriers. The US scheduled airline industry improved its fuel productivity steadily from 1.11 RTK per litre ($= 2.9$ RTM $\times 0.3831$) in 1978 to 1.42 RTK (3.7 RTM) in 1988. As in Canada, most of this 30% increase in fuel productivity occurred within five years after US deregulation. Canada had marginally higher fuel productivity relative to the US.

Proper measurement of capital productivity requires substantial data and research time beyond the scope of this chapter. However, aircraft utilization is used here to infer general directions of changes in the productivity of capital input. Two major indicators for aircraft utilization are passenger load factor and hours flown per aircraft per year.

Passenger load factor in Canada improved substantially from roughly 63% in the 1978–82 period to 65–69% in the 1983–89 period (Figure 5.11). Although the improved state of the economy may have partly contributed, regulatory relaxation and deregulation are main candidates for the improved load factor. It is worth noticing that airlines in Canada experienced a lower load factor (63%) during the 1978–80 economic boom than the 1981–83 economic recession. Furthermore, the load factor improved by 2 to 3 percentage points since 1988, the year in which deregulation took effect.

The US scheduled airline industry improved their passenger load factors dramatically after 1978. Their passenger load factors ranged from 49–56% during the 1970–77 period but jumped to 61.5% in 1978. Since 1978, their load factor has stayed in the 59–63% range. Historically, Canadian level I carriers have always enjoyed a higher load factor relative to the US (higher by a factor of more than 5% prior to US deregulation). This difference is attributable, at least partly, to the difference in airline regulations: the US CAB did not regulate airline capacity while Canada's CTC regulated capacity as well as entry/exit and prices.

The statistics on revenue hours flown per aircraft for the level I carriers in

Canada also improved, from 2,059 hours per aircraft-year in 1983 to 3,165 hours in 1988, a 54% increase (Figure 5.12). The competitive pressure for efficiency improvement intensified by regulatory relaxation and deregulation as well as the general improvement of the economy, appear to have contributed to the improved utilization of aircraft. It is noteworthy that the hours flown per aircraft in 1988 is also substantially higher (3,165 per aircraft) than that of the previous peak utilization period which occurred in 1979 (3,034 per aircraft).

The above evidence of various measures of partial factor productivities (labour, capital and fuel) indicates that the Canadian airline industry appears to have improved its productivity since the regulatory relaxation in 1984 and deregulation in 1988. A formal computation and analysis of total factor productivity (TFP) is required to make a firm inference on the issue. However, this cannot be done here because measurement of TFP requires far more detailed data on revenues, outputs, operating statistics, capital investments, and other cost elements than those currently available to us.

Welfare gains
Welfare gains from the regulatory changes (relaxation and deregulation) consist of the changes in consumer and producer surpluses. Changes in quality of service must be taken into account in computing consumer surplus. Ideally, the actual values of prices, traffic volumes, and service quality measures under a deregulated situation should be compared with the respective values which would have prevailed under the counter-factual scenario – the regulated case. However, the data available to us does not allow such a rigorous measurement. Therefore, in this section we attempt to apply the empirical results of Gillen *et al.* (1987: 130–4) to the 1988 factual data for the Canadian airline industry in order to make some inference about the effects of regulatory relaxation, deregulation, and privatisation of Air Canada on consumer and producer surpluses.

Effect of reduced airfares on consumer surplus: The reduced airfares, and increased availability of various discount fares have no doubt increased the consumer surplus of the travelling public substantially. Lower fares increase consumer welfare in two ways: by stimulating the market and thus increasing overall travel volume; and by increasing the size of consumer surplus for the individuals who would have travelled at higher regulated fares. Between 1984 and 1988, the average yield per passenger-kilometre increased by only 5% (decreased by 10% in real terms) as compared to a 60% increase (4% reduction in real terms) during the 1978–84 period. Without a formal analysis, it is impossible to discern the effect of the regulatory and ownership changes on fares from the effects of other changes such as airline input prices, traffic composition, average distance of travel, etc. On the basis of an in-depth analysis of the productive efficiency and structure of the airline industry, Gillen *et al.* (1987) concluded that deregulation and privatisation would reduce average airfares by 8.8% and 4.6%, respectively. These results were applied to

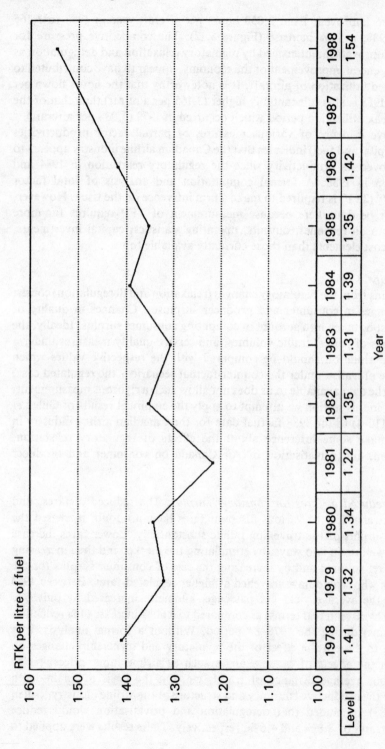

RTK per litre of fuel

	1978	1979	1980	1981	1982	1983	1984	1985	1986	1987	1988
Level I	1.41	1.32	1.34	1.22	1.35	1.31	1.39	1.35	1.42		1.54

Year

—■— Level I

Source: Compiled from Statistics Canada 51–002, *Air Carrier Operations in Canada;*

Figure 5.10   RTK per litre of fuel: level I carriers.

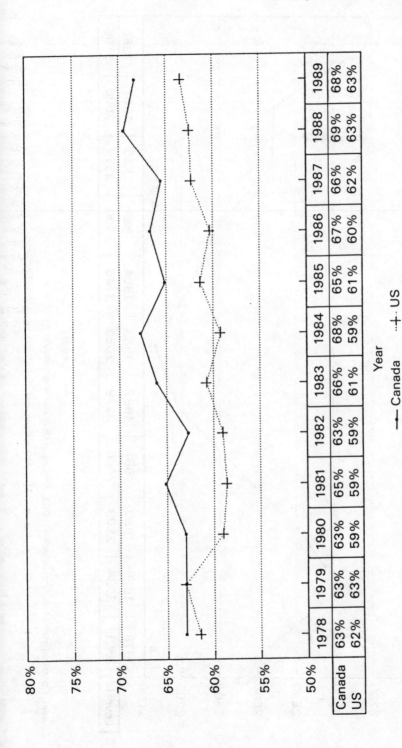

	1978	1979	1980	1981	1982	1983	1984	1985	1986	1987	1988	1989
Canada	63%	63%	63%	65%	63%	66%	68%	65%	67%	66%	69%	68%
US	62%	63%	59%	59%	59%	61%	59%	61%	60%	62%	63%	63%

Year

—•— Canada   —+— US

Source: Compiled from Statistics Canada 51–002, *Air Carrier Operations in Canada*, Air Transport Association of America *Annual Report* (various issues).

Figure 5.11   Passenger load factor: Canadian level I and US scheduled carriers.

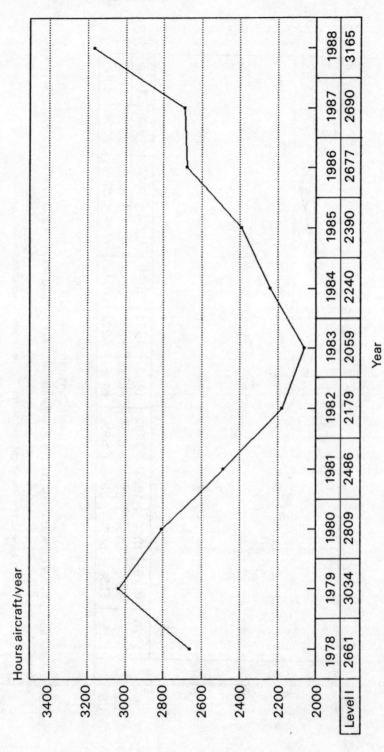

Hours aircraft/year

	1978	1979	1980	1981	1982	1983	1984	1985	1986	1987	1988
Level I	2661	3034	2809	2486	2179	2059	2240	2390	2677	2690	3165

Year

— Level I

Source: Compiled from Statistics Canada 51–002, *Air Carrier Operations in Canada;* 51–206.

Figure 5.12   Hours flown per aircraft per year: level I carriers.

Table 5.8 Changes in consumer surplus (1988).

	Factual cases	Counter-factual cases without deregulation	Counter-factual cases without deregulation or privitization
Price reduction	—	− 8.8%	− 13.4%
Price (cents/RTK)	C¢111.4	C¢122.2 (= 111.4/(1 − 0.088))	C¢128.6 (= 111.4/(1 − 0.134))
Volume reduction	—	9.7%	14.7%
Traffic volume (million RTK)	5,961	5,434 (= 5961/1.097)	5,197 (= 5961/1.147)
Approximate consumer surplus gains (millions)	—	$632.4	$959.6
Total revenue (billions)	$6,642	—	—
Percentage of consumer surplus gains to total revenue	—	9.52%	14.45%

1988 factual data in order to estimate the reduced consumer surplus which would have prevailed if deregulation and/or privatisation were not in place in 1988. This calculation is reported in Table 5.8.[51]

The 1988 average yield for the Canadian airline industry was 111.4 cents per RTK and total volume of traffic was 5.961 billion RTK. According to the results of Gillen *et al.* (1987), the average yield would have been 128.6 cents without deregulation and privatisation, and 122.2 cents without deregulation (but with privatisation). Application of the price-elasticity of demand, − 1.1[52] implies that traffic volume would have been reduced from 5.961 to 5.434 billion RTK (8.8% lower) in the absence of deregulation, and to 5.197 (12.8% lower) in the absence of deregulation and privatisation. The actual traffic increased by about 33% between 1984 and 1988, a part of which is no doubt natural growth due to growth of the economy. Using the volumes and prices which would have prevailed in the absence of deregulation and/or privatisation, it is possible to linearly approximate the difference in consumer surplus between the 1988 factual results and the counter-factual situations without deregulation and/or privatisation. The results show that deregulation and privatisation have increased consumer surplus in 1988 as follows: deregulation: $632 million (or 9.32% of the industry revenue); deregulation and privatisation: $960 million (or 14.5% of the industry revenue).

Effect of quality of service on consumer surplus: According to the empirical results of Morrison and Winston (1986), most of the consumer surplus gain for short-distance (less than 1,000 miles) travellers due to US deregulation was attributable to the increased schedule frequency while most of the consumer surplus gains for long-distance travellers came from reduced fares. In Canada, between 1984 and 1989 departure frequency increased by 114% (10% increase in jet, and 168% increase in non-jet) in the North, and by 100% in the South.

There is no doubt that this tremendous increase in departure frequency has increased consumer welfare substantially. However, we are unable to quantify this welfare gain due to the lack of data by fareclass and by origin–destination pairs.

Producer surplus: When computing the change in consumer surplus caused by price changes it is assumed that all of the efficiency gain due to the regulatory and ownership changes are being passed onto travellers. From the earlier discussion on airline profitability, it is noted that the Canadian airline industry's total profits in recent years exceed those of the late 1970s. However, the ratios of operating profit and of net profit to total revenue were reduced in recent years as compared to the late 1970s. Therefore, at the present we observe no sign of increased producer surplus per unit of output after regulatory relaxation and deregulation. This suggests that cost savings due to improved efficiency have been passed onto consumers.

Canadian carriers' performance in international markets

The airline business in transborder and international markets is still regulated by bilateral treaties between Canada and foreign countries. As such, deregulation of the domestic market only indirectly influences carrier performance in international markets. From 1978 to 1984 Canadian carriers lost market share to US carriers on the transborder routes as the US carriers become aggressive and efficient. Canadian carriers also lost some market shares on other international routes from 1978 to 1984. Post-1984 data is not available.

5.7 Privatising Air Canada

On 12 April 1988, less than two weeks after he became the Minister responsible for privatisation, Don Mazankowski (the former Minister of Transport) announced that all of Air Canada would be sold to the public as 'market conditions permit', with an initial treasury issue of up to 45% of the shares. With assets of $3.18 billion and revenues of $3.13 billion in 1987, the sale of Air Canada represented by far the most ambitious privatisation by the Mulroney Government (see Stanbury, 1988).

Professor John Baldwin (1985: 141–2), a long time observer of Air Canada, suggested that the sale of Crown operations may have several objectives:

- to improve efficiency perhaps due to constraints imposed by government ownership, e.g. for capital of expansion or by the lack of external stimuli;
- to contribute to the 'capitalist ethic' through widespread share ownership;
- to reduce the strain upon the treasury where government ownership has been used to rescue dying industries;

- to enforce on government enterprises an appropriate cost of capital; and
- as a prerequisite for a change in the degree of competition in the industry, e.g. deregulation.

The Economic Council of Canada (1986: 67) recommended that Air Canada be privatised after the National Transportation Act was enacted to 'ensure that the gains from a deregulated airline industry are fully realized'. Its principal arguments supporting the privatisation recommendation[53] were first, Air Canada could have an unfair advantage over its rivals because of its access or potential access to the public purse. It is very difficult to ensure, for example, that a Crown corporation operates subject to the constraint of the appropriate cost of capital. Second, there is considerable doubt whether policy makers are able to 'detach themselves sufficiently from a public corporation to accept the results of the competitive process'. They 'will inevitably have some involvement in major corporate decisions'. Third, public ownership can inhibit management's ability to undertake initiatives designed to enhance the corporation's performance. Air Canada may not be sufficiently flexible under government ownership to adapt to stronger competitive pressures under deregulation. Fourth, 'privatization will not lead to the sacrifice of important public-policy objectives. The primary loss would be felt by those who place a high value on the symbol of a government-owned airline' (Economic Council of Canada, 1986: 66).

Immediately after the Minister's announcement that Air Canada would be privatised, opposition Leader John Turner called the move 'a clear breach of faith between the Prime Minister and the Canadian people'. NDP leader Ed Broadbent described it as 'the triumph of Conservative ideology over good, practical, Canadian common sense' and vowed his party would 'do everything we can to stop the Government'.[54] The president of the airline division of CUPE said privatisation will lead to a deterioration in labour relations and a loss of jobs and threaten the company's pension plan: He said, 'What the Government fails to realize is that the people of Canada already own Air Canada'.[55] An Angus Reid-Southam News poll conducted in March 1988 indicated that 53% of Canadians thought Air Canada should not be sold, while 35% favour privatisation.[56]

The chairman of Air Canada, Claude Taylor, called the Minister's announcement 'a giant step towards the realization of a dream I have cherished a long time'. President Pierre Jeanniot described privatisation as the 'financial key to Air Canada's future'. It will allow 'the airline to plan for fleet expansion and renewal as well as pursue new business opportunities'. The chairman of the Air Canada Employee Ownership Committee, which had 7,500 members shortly after it was formed in 1985, said, 'I don't think we're going to have any trouble getting people enthused [sic] again'.[57]

Editorially, the influential *Globe and Mail* supported the proposal, but noted, however, that the credibility of the government's promise not to interfere with the commercial decisions of the airline after its stake is reduced to 55% 'will be critical to buyers of the minority stake'.[58]

The privatisation of Air Canada was made subject to a number of conditions, most of which were enshrined in the enabling legislation passed in August 1988.

● The headquarters must remain in Montreal.
● The airline must, for the indefinite future, maintain its major operational and overhaul centres in Winnipeg, Montreal and Toronto.[59]
● No more than 45% of the share would be sold in the initial offering and the proceeds would go to the airline, not the federal government as owner.
● Employees had to be given the first chance to buy shares through a payroll deduction scheme or similar programme. Small shareholders were next on the list of preferred buyers, followed by institutional investors, and finally by foreigners.
● The Government's 55% stake will be voted in accordance with the majority of the new private sector shareholders so that there will be 'a clear arms-length relationship'.
● No individual shareholder is allowed to hold more than 10% of the shares sold to private investors (hence 4.5% of the initial 45% offering).
● Total foreign ownership is limited to 25% (or 11% of the initial 45% offering).

Note that the 10% limit on individual shareholdings constitutes a serious constraint on the 'revenge of the capital market' in the event that management performs poorly. The federal legislation followed the unfortunate Alberta precedent when Pacific Western Airlines were privatised in December 1983. The legislation provided that no individual or *group of associated individuals* could vote more than 4% of the stock.[60] Such a constraint gives management effective control – even though it owns no shares at all. Hence, Canada's two largest airlines are now privately-owned, but effectively controlled by management.

On 19 May 1988, Bill C-129, the Air Canada Public Participation Act was given its first reading in the House of Commons, despite procedural manoeuvring by the Opposition. The second reading debate began on 24 May. Both the Liberals (including Lloyd Axworthy) and NDP strongly denounced the bill.[61] Bill C-129 passed the second reading by a vote of 104 to 38 on 7 June. A Commons committee held hearings beginning 14 June and referred the bill back to the House with only one minor amendment on 23 June. Bill C-129 was passed by the Commons on 18 July and by the Senate, after only three days of Committee hearings, on 17 August. It was given Royal Assent the next day.

On 25 August 1988 Air Canada filed its preliminary prospectus in Quebec for its first privatisation issue. The shares were eligible for the provincially-subsidized Quebec Stock Savings Plan. Two days later Air Canada published full-page ads stating in many newspapers, 'History is now being made' and including an 'expression of interest' form for potential investors. However, on 29 August the Liberal Party's transportation critic announced that if 45% of Air Canada is sold and the Liberals form the next government (an election was expected in the fall), the balance of shares will *not* be sold.[62] Air Canada began distribution of 650,000 copies of its prospectus.

On 26 September, Air Canada filed its final prospectus which stated that net income after taxes was $101 million for the year ended 31 March 1988. The next day, the price of Air Canada shares was set at $8 for the 30.8 million shares to be issued (42.8%). In addition, brokers had an option to buy another 3.5 million shares at $8. On 29 September the *Financial Post* (p. 1) reported that 'demand for Air Canada shares is so strong that virtually none will be allocated to foreigners, and a thriving secondary market is already springing up across the country'. A Canadian dealer said, 'Air Canada is cheap as dirt, compared to any US airline stocks. There's no doubt it will trade in the $9-$10 range at least'. The *Globe and Mail* (p. B1) reported that Air Canada will net only $225.8 million out of the sale of its shares for $246.2 million. Underwriting fees were $12.3 million and the airline absorbed $8 million on the shares sold to employees at a discount. Some 80% of employees bought shares in the first issue.

Air Canada shares were listed on five Canadian stock exchanges on 13 October 1988. They opened at $8.25 on the Montreal Stock Exchange. On 28 October, the *Globe and Mail* (p. B4) reported that underwriters would not be exercising their option to buy an additional 3.5 million shares of Air Canada at $8. Air Canada shares closed on the TSE at $7.60. At the end of March 1989 Air Canada's stock was trading at $11.75, up from $8 in mid-January when PWA Corp announced its agreement to acquire Wardair for $300 million.[63] In 1988 Air Canada had a net income of $96 million ($2/share) on revenues of $3.43 billion.

On 31 May, the *Globe and Mail* (pp. B1, B4) reported that Air Canada indicated that holdings of its shares by foreigners were near the 25% limit set in Bill C-129. It instructed its transfer agent to refuse to transfer shares to a non-resident if it would breach the limit. On 12 September 1989 the *Globe and Mail* (p. B25) reported that Air Canada stated that foreigners had hit the 25% limit on ownership of its stock. It stopped transfers of the stock from residents to non-residents. Air Canada has some 100,000 stockholders of which 18,000 are company employees. The stock hit a high of $14.83 in August 1989.[64]

Air Canada issued its final prospectus on the second part of its privatisation on 6 July 1989. It proposed to sell 41.1 million shares at $12 for 57% of the equity (vs. 30.8 million shares at $8 in September 1988). This time, the proceeds went to the federal government which created the Crown corporation in 1937. The Liberals' privatisation critic said the sale was timed to occur when Parliament was not sitting to avoid questions by the Opposition. The Minister responsible for privatisation said the price of $12 represented a good return to the government, although in May he hinted the price might be as high as $16.[65]

Unlike the first tranche, this time institutional interest was strong – in part because of PWA Corp's takeover of Wardair which was completed in April. As one analyst put it, 'It's become a very powerful oligopoly'. Julius Maldutis of Salomon Brothers of New York stated: 'I believe that at $12, Air Canada is highly undervalued and is an attractive long term investment'.[66] By the end of the first week after the second tranche was sold, Air Canada's shares had climbed to $12.75.

While the president and chairman of Air Canada began pushing for its privatisation in 1984 and 1985, the key player was Don Mazankowski. As Mulroney's first Minister of Transport he pushed through deregulation of the transport sector (although the process was irrevocably begun by the Liberals' Lloyd Axworthy in May 1984). He was also strongly committed to privatising Air Canada; he began studies on the issue within weeks of the Tories assuming office in 1984. He even persisted after the Prime Minister said Air Canada was not for sale in January 1985.[67] He also persisted despite Mulroney's decision in July 1987 to block the sale of all of Air Canada in one issue.[68] This meant that the privatisation had to be done *after* the sharp drop in world stock markets in October 1987. It seems reasonable, in light of Air Canada's 'pressure' in January 1988,[69] to believe that Mazankowski made the sale of Air Canada a condition of taking over the privatisation portfolio on 31 March 1988. Besides, the PM 'owed him one'. 'Maz' had been a vitally important minister in the Tory cabinet. Mazankowski moved very quickly to irrevocably commit the government to this controversial move.

5.8 Future developments

Industry evolution

In the next few years the Canadian airline industry will experience a rationalization of capacity. The shake out in the charter airline segment has already been discussed. Excess capacity has also been plaguing the scheduled segment of the Canadian industry.[70] We have argued that Canadian markets are unable to support more than two competitors of minimum efficient size. When Wardair entered the scheduled market in 1986 in competition with Air Canada and CAIL, the consequence was an introduction of excess capacity into the market.[71] In early 1989 CAIL acquired Wardair, but it still had Wardair's aircraft, notably its new fleet of 12 A-310-300s. In order to cover the corresponding overhead costs it had to keep these aircraft flying. In spite of the elimination of a primary competitor, airfares fell during the second quarter of 1989.[72] This was a direct consequence of the excess capacity. In early 1990 CAIL announced that it was going to sell the entire Wardair fleet in order to reduce capacity and hopefully return to profitability.[73]

It seems likely that the industry will see the emergence of a price leader. Already, CAIL has attempted to raise prices but Air Canada has refused to go along. Whenever Air Canada raises its prices CAIL quickly matches.

Another development which is likely to occur has to do with the nature of the relationship between feeder and trunk carriers. In the early days of deregulation, feeders had excellent growth prospects, as trunks were turning over low traffic points to them. However, as the feeders build traffic due to their more frequent and well-timed services, the trunks will eventually reclaim some markets. A feeder carrier will be unable to redeploy its equipment to other geographic areas, as the trunk has other affiliates. The result is likely to be (a) a maturing of the feeder carrier industry, and (b) the trunks acquiring majority ownership of feeder carriers in order to have the flexibility to put their jets into developed markets. InterCanadian found these choices unacceptable, and decided to defect from its trunk and strike out on its own. Only time will tell if this will prove to be a viable alternative for a feeder.

A final issue is whether or not alliances will be built between US and Canadian carriers. There have been some market by market cases of Canadian and US carriers feeding each other, but no systematic policy between a US carrier and a Canadian carrier to mutually feed each other. Similarly, the role of the Canadian airlines in a globalizing airline industry remains to be seen.

Required changes in government policy in domestic markets

As the Bureau of Competition Policy noted on 24 April 1989 in regard to the takeover of Wardair, 'The domestic scheduled airline industry exhibits a number of entry barriers which significantly hinder the prospect for entry by new firms in the short to medium term.' The Bureau referred to two regulatory barriers: capacity at Toronto's Pearson airport, and the 25% limit on foreign ownership. The Bureau also pointed out that cabotage, the operation of point-to-point scheduled services within Canada by foreign carriers, is prohibited by federal law (as it is in the US). While the opening of Terminal 3 at Toronto's Pearson airport in 1990 will provide more gates and ticketing space, it will not solve the problems engendered by the federal government's restrictions on the number of takeoffs and landings on the two runways.

In any event, government policy should consist of the following: (1) close scrutiny of the pricing behaviour and other forms of co-ordination between Air Canada and CAIL under the Competition Act (Stanbury, 1986); (2) elimination of the 25% maximum foreign ownership constraint where firms are created or acquired with a view to challenging the duopoly; (3) unilaterally permitting cabotage in Canada on a limited basis where domestic fares and service do not reflect competitive behaviour.

International policy

Canada's current policy for international air services was developed primarily in the regulated era. This included a divide-the-world policy whereby spheres of influence were assigned to Air Canada and CP Air/CAIL. This policy has been revised somewhat in the last few years, but with the exceptions of London and Paris, no dual designations of international routes have been achieved. Of vital interest in the future will be developments in the US–Canada air bilateral. Dresner *et al.* (1988) discussed the evolution and current status of this critical treaty. Perhaps somewhat surprisingly Canada has proposed complete deregulation of North American air transportation, including cabotage rights. The US rejected this position, favouring instead deregulation only of routes which actually transit the border. Tretheway (1990a) discusses what globalization of air transport means and its current prospects. Should the industry in fact globalize, Canada will need to develop an appropriate policy.[74]

5.9 Lessons for other countries

Evolution versus revolution

While the US took a 'sudden sharp change' approach to deregulating its airline industry, Canada took an evolutionary one. As a result, almost all of the structural and efficiency-promoting adjustments took place between May 1984 – when regulation was greatly liberalized – and full, formal deregulation in southern Canada on 1 January 1988. In 1984, the carriers, the interested public and government officials assumed that full deregulation was inevitable. This view was greatly reinforced in mid-1985 when the newly-elected Tory government published its policy paper *Freedom to Move* which proposed complete deregulation. Because of the gradual nature of the process, the incumbent Canadian carriers thus had time to adjust and to erect entry barriers. The result, thus far, has been that no new (successful) entry has taken place.

While Canada took an evolutionary approach, in some ways it had further to travel than the US airline industry for two reasons. First, in late 1983, three of the four regional carriers (PWA, Quebecair and Nordair) were owned by the governments of Alberta, Quebec and Canada respectively. Moreover, the largest carrier (Air Canada) – with over one-half the domestic market – was also owned by the federal government. A little over six years later all four were privately owned. Second, the policy and regulatory regime in place in Canada in the early 1980s was more elaborate and more restrictive than that in the US. Carriers were divided into three levels and their roles and responsibilities set out in federal policies designed to limit competition between and among them. The CTC imposed extra constraints on carriers, e.g. limiting the equipment some could use, specifying frequencies, and imposing stops on some routes. There was no counterpart in Canada to the much less restrictive regime for US commuter carriers.

Control of up/down stream markets

The alliance of feeder carriers with trunk carriers has created a very effective entry barrier for the trunk airline industry. Although feeder traffic may account for only 5% of a carrier's total traffic, it may well have a great effect on its profits at the margin. The trunk's point to point travellers cover the cost of the flight, whereas the feeder passenger revenues are largely profit. Further, a lower portion of feeder passengers obtain discounts, so their yields are much higher. With the trunk duopolists already controlling virtually the entire feeder market, a new entrant would be at a significant disadvantage. A related entry barrier is control of the distribution channel, by the CRS system. Not only does this influence the travel agent's choice on behalf of the consumer as to which airline will be chosen, control of the CRS system generates highly valuable market information.

In our view, countries contemplating regulatory reform should seriously consider preventing trunk carriers from controlling feeder carriers and computer reservation systems.

Infrastructure constraints and effects on entry/competition

Canada, as with much of the world, is facing growing shortages of capacity on both the ground and air side at airports. Control of these scarce resources also acts as a barrier to entry. It remains to be seen whether Intair will be able to successfully compete in the congested Toronto–Montreal–Ottawa corridor.

Role of government ownership of carriers

As noted under 'Evolution versus revolution', government ownership of four of the six largest scheduled carriers in 1983 did not inhibit the process of liberalization and ultimately deregulation. However, privatisation was eventually seen as necessary to ensure that the carriers were free to respond to market forces under deregulation, and thus allow the reaping of the full benefits of deregulation. Without privatisation, deregulation would not be able to achieve its full potential benefits of greater efficiency and optimal service levels and patterns. There was far less political controversy over privatisation than there was over deregulation. Partly this was due to a general climate of public opinion favourable to privatisation. In the case of Air Canada, the need to obtain equity capital to finance fleet renewal in the presence of strong competition was a major factor. However, Air Canada's long history as a 'favourite child' Crown corporation (1937–87) has left it with substantial advantages over its duopoly rival, e.g. a more extensive set of international routes, particularly those to the US.

Relationship between deregulation and privatisation

In our view, deregulation is much more important than privatisation in terms of improving economic efficiency (static and dynamic) in the airline industry. This is consistent with previous studies of the Australian airline industry which found that the injection of competition greatly improved the performance of the Crown-owned carrier. To get the full benefits of deregulation it is necessary to privatise government-owned carriers for several reasons. First, it avoids the possibility of implicit subsidies, such as cheap debt capital, which would result in unfair competition. Second, privatisation permits the capital market to extract its revenge (a hostile tender offer) where management has performed poorly. Poor performance by a government-owned carrier is less likely to result in a change in managerial practices, provided political criteria are met. Third, privatisation usually makes it easier for the airline to adjust to the more intense competition associated with deregulation, e.g. renegotiating union contracts; changing route structure; altering the fleet. In general terms, the carrier must reduce its costs and become more customer-oriented. Continued government ownership of a carrier could result in political pressures being put on it to provide uneconomic services, thus undermining its ability to compete (assuming no subsidy) in markets which are now competitive.

It should also be noted that privatisation which limits private-sector shareholders, individually or in groups, to a small fraction of total voting shares (as was done in the case of PWA – now CAIL, and Air Canada) is undesirable. It ensures that management will effectively control the airline, subject only to pressures from bankers if it fails to meet its short and long-term debts.

Domestic policy in context of evolving globalized industry

An important issue which needs to be resolved is what is the appropriate policy for a country if the industry globalizes. One view in Canada is that our two duopolists should merge together to form a monopoly that is of sufficient size to compete with the other mega-carriers. An alternative view is that the duopolists should be permitted to align separately with these mega-carriers. In this way, there will be competition in Canadian domestic markets: between the global carriers. Regulations limiting foreigners' ability to own Canadian domiciled carriers might prevent our carriers from fully participating in the globalization of the industry.

Role of domestic ownership requirements

The 25% limit on foreign ownership of domestic air carriers in Canada provision in the National Transportation Act of 1987 was said to have been put in to mirror that of the US. In any event, it is unnecessary from the perspective of ensuring that Canadian authorities have sufficient control over domestic carriers. The authority in the National Transportation Act is sufficient, regardless of which nationalities the owners are. Foreign ownership provisions are relatively more burdensome for small countries like Canada than for large ones like the US. In light of the Canada-US Free Trade Agreement (which came into effect 1 January 1989), the ownership constraints are an anomaly.

Regulating brand loyalty creating gimmicks

Frequent flyer programmes were shown to act as an entry barrier. They are very effective in generating product loyalty among consumers, but in the long run, if consumers pay higher prices due to the lack of competition, there may not be a net gain. Countries which have not allowed these programmes should give very serious consideration to the anti-competitive impacts of allowing their introduction.

Consumer protection

Canada's National Transportation Act, 1987 did not have any consumer protection provisions. One view of consumer protection is that information is costly and therefore some degree of regulation is required in order to achieve the economically optimal result. The US, for example, has required carriers to report on time performance statistics so consumers can make more informed choices. Another aspect of consumer protection has to do with ticket prepayment. Air transport, unlike many retail industries, requires advance payment for tickets. In industrial transactions which require prepayment (e.g., purchase of aircraft), customer deposits are protected by detailed provisions in the sales contract. Such contracts are not feasible in this retail industry, justifying some form of regular protection for the consumer. The recent bankruptcies of several Canadian charter airlines has resulted in substantial losses to consumers as their prepaid tickets were not protected in trust funds. To some extent this was a failure of enforcement rather than policy, although the existing law leaves it to provinces to set policies and many of them have not.

Importance of a well-prepared competition authority as the regulatory agency is phased out

Canada did not do a good job in shifting the mode of social control over the domestic airline industry from direct regulation to competition policy. The 'regulated conduct' exemption under the Competition Act is not defined in the statute; it rests on rather ambiguous case law.[75] The situation is further complicated because fares (and entry) continue to be regulated in the north but not in southern Canada. Yet many of the same carriers serve both regions. In any event, deregulation should be preceded by a clear announcement indicating when general competition law rules will apply to a previously regulated industry. This may require some amendments to clarify the relationship between the two policies.

Continued regulation of remote community markets

The US chose to deregulate service to all communities, even those in small isolated Alaskan regions. In order to achieve the social policy objective of maintaining services to some communities, even where uneconomic, the US introduced a subsidy system. Canada has chosen to continue to regulate service to remote northern communities. However it has reversed the onus of the burden of proof for regulatory hearings, so that incumbent airlines must make the case if they wish to block new entrants. The result to date in Canada has been an introduction of new services in many of these remote regions, and at least some availability of discount airfares.

Importance of public access to information

The Canadian government does not publish information on individual air carriers, and what information it does publish is provided with a lag of roughly two years. It is not clear who this protects. The two duopolists certainly know what is going on in their markets. The government receives information from the carriers on a confidential basis. Only consumers and academic researchers are denied access to the information. Since any regulatory reform will have its uncertainties, it is essential to have information available so that performance in markets can be continuously monitored: not just by the government but by independent parties as well. Both the industry and the consumer would be well served by low cost public access to timely information.

Notes

1. A more detailed account is found in Gillen *et al.* (1987).
2. Canada, House of Commons, *Debates*, 1938, p. 1554.
3. Canada, House of Commons, *Debates*, 1943, p. 1778.
4. From 1938 to 1944 the airline industry was regulated by the Board of Transport Commissioners. From 1944 to 1967 regulation was exercised by the Air Transport Board. In 1967 the Air Transport Committee of the Canadian Transport Commission took over until 1 January 1988 when the National Transportation Agency was established.
5. In 1977 the number was reduced to four when Pacific Western Airlines took over the failing Transair which was based in Winnipeg.
6. Routes from Canada to the US are referred to as transborder routes.
7. The policy was reiterated in another statement by the Minister of Transport on 23 November 1973.
8. The Air Transport Board held that CP Air should not operate 'bush type' services and stated that the airline's future lay in the provision of scheduled service. According to Baldwin (1975: 44), since it was 'the only carrier designated for expansion of scheduled services, it was implicitly being promised a protected position'.

9. Both Air Canada and CP Air were permitted to offer a very few seats at a discount in 1977 and 1978.
10. See Reschenthaler & Stanbury (1982), (1983).
11. The CTC also did not try to stop PWA Corp's acquisition of CP Air in 1987 (for $300 million) or its purchase of Wardair (then a failing firm) in 1989 (for $250 million).
12. Generally, see Bailey *et al.* (1985), Breyer & Stein (1982), and Derthick & Quirk (1985).
13. Air Transport Association, *Air Transport*, Washington D.C., various annual issues.
14. See Kraft *et al.* (1986) for a discussion of the history and concept of airline seat management.
15. From 1962 to 1974, 30% of US communities lost their air service in spite of government regulation of the industry (Jordan, 1970). In many of these communities, as well as in communities receiving regulated air transport, commuter airlines, operating aircraft too small to qualify for CAB regulation, were appearing. Federal Express is another example of how a carrier operating outside CAB regulation was able to tap into a previously unserved air transportation market. More generally, see Derthick & Quirk (1985).
16. In the early 1960s Wardair began offering charter flights to Europe at fares far below those of the scheduled carriers. It grew rapidly. Later Wardair offered charter flights to 'sun spot' locations in the US and elsewhere. These too were very popular. See Reschenthaler & Stanbury (1982).
17. The minority Tory government of Joe Clark, elected in April 1979, had been defeated December 1979. The Liberals obtained a majority in the general election in February 1980, but did not assume office until early March.
18. Such fares had to be justified, not simply filed; they could apply only to round trip travel; reservations had to be made 14 days in advance; and the minimum stay must include the first Saturday following the departure (Reschenthaler & Stanbury, 1983: 215–16).
19. See *Globe and Mail*, 9 April 1984, p. 2.
20. The *Interim Report of the Air Transport Committee of the Canadian Transport Commission on Domestic Charters and Air Fare Issues* (see Canadian Transport Commission, 1984) agreed with the Minister of Transport that relaxation of regulations is necessary, but does not advocate complete deregulation. (Note that while this document was *dated* 9 May it was not released until *after* the Minister's policy statement on 10 May.)
21. Generally, see Oum & Tretheway (1984), Canada, Department of Transport (1984), and Gillen *et al.* (1987).
22. See Canada, Department of Transport (1985) and Canada, House of Commons Standing Committee on Transport (1985).
23. For more detail, see Stanbury & Tretheway (1987).
24. The South is defined roughly as the area below a line beginning with the 50th parallel on the Atlantic coast to the Manitoba-Ontario border, then northwesterly to the 53rd parallel at the Manitoba-Saskatchewan border, the northwesterly from that point to the 55th parallel at the border of Saskatchewan and Alberta, then along the 55th parallel to the Pacific coast, see Figure 5.1. The Minister of Transport has the power to designate any point in the South as being part of the regulated area.
25. See, for example, studies of international airlines in Tretheway (1984), and Caves *et al.* (1987).
26. Using the results of Carlton *et al.* (1980), the value of an on-line connection to travellers can be estimated to be about $31 (1989 Canadian dollars).
27. See Tretheway (1989) for a discussion of the potential anti-competitive effects of frequent flyer programmes.
28. See Gillen *et al.* (1985), especially chapter 8.

29. CP Air announced it had reached an agreement to acquire EPA for $20 million in April 1984.
30. See the detailed discussions in Gillen *et al.* (1987).
31. The deal, for $300 million, was announced on 2 December 1987. At that time, PWA Corp had revenues of $352 million versus $1400 million for CP Air. The deal was closed on 30 January 1987 and on 24 March the new corporate identity, CAIL, was announced.
32. It should be noted that Quebecair was not folded into the CAIL operation. Quebecair was merged into a feeder carrier, InterCanadian, which was 35% controlled by Nordair (and later by PWA Corporation). InterCanadian eventually became Intair. See section 'Consolidation of the feeder carrier network'.
33. On 24 April 1989 the Bureau of Competition Policy confirmed that it would not be challenging PWAC's takeover of Wardair under the Competition Act. The Director of Investigation and Research said, 'although potential bidders have expressed some interest in acquiring Wardair, no firm written offers have been brought forward' despite the extensive publicity following the 19 January announcement and the companies' 6 April announcement. In a much-longer-than-usual press release, the Bureau stated that 'The acquisition of Wardair by PWAC will result in the removal of an effective and vigorous competitor'. The Bureau noted that in most domestic markets there will be a duopoly. 'The stock market reaction, which led to higher stock prices for both Air Canada and PWAC, suggests that investors expect increased profits for both rivals of Wardair'. The key reason for not challenging the takeover was the fact that Wardair was a failing firm. The Bureau stated that 'a firm that is facing certain and imminent financial failure will cease to exercise any competitive influence in the market after its failure. Therefore, the loss of its influence on the marketplace cannot be attributed to the merger'. The Bureau concluded that Wardair was likely to fail: it did not have the cash to meet a large payment due in the fall of 1989. 'There were no alternatives to the merger that would result in a more competitive environment'.
34. Air Canada's market share is 55% and CAIL's is 44%, according to Cecil Foster, 'Hard landing for airlines' earnings', *Financial Post*, 26 February 1990, p. 3.
35. 'Air Canada Cutting 460 Jobs in Wake of Low 1989 Profit', *Globe and Mail*, 2 March 1989
36. 'Wardair cited in PWA loss of 56 million dollars', Vancouver *Sun*, 1 March 1990, p. D4.
37. City Express is a small, independent commuter carrier. Its operations use the very small and short runway at Toronto Island airport near the downtown. However, that airport is being expanded somewhat. In 1990 Air Ontario began using it. Newspaper reports suggest that City Express may be for sale. See *Globe and Mail*, 26 May 1990, p. B3.
38. See Z. Cashmeri & D. McArthur, 'Air Toronto purchased by Air Canada', *Globe and Mail*, 6 April 1990, pp. B1, B4.
39. See C. Foster 'Wardair to pay commuter fares for some connecting passengers', *Globe and Mail*, 18 January 1989, p. B10.
40. The 20% of agents who are not automated account for a very small proportion of airline ticket sales.
41. For example, the agent will have greater confidence that the information in the CRS is most up-to-date for the owner airline than for other airlines. This is especially important when booking last-minute and usually full fare tickets.
42. Competition Tribunal, Consent Order and Reasons for Consent Order (Ottawa, 17 July 1989) re Director of Investigation and Research and Air Canada, PWA Corporation *et al.*
43. See C. Foster 'Air Canada fights back at charters', *Financial Post*, 15 January 1990, p. 15.

44. On 28 April 1990 the *Globe and Mail* (pp. B1, B8) reported that Soundair Corp., owner of chartered carrier Odyssey International (five planes), was placed in receivership by the Royal Bank to try to recover over $500 million in loans. Thomson Vacations, which provided 75% of Odyssey's passengers, went into receivership the day before. Odyssey was one of five charter carrier to fold in 1990 (Crownair; Ports of Call; Vacationair; Holdair). Air Toronto, owned by Soundair, continued its scheduled service. The bankruptcies pointed out a loophole in the 1988 National Transportation Act regulations, namely that a charter airline run by a scheduled passenger carrier is exempted from either obtaining a performance bond or a guarantee from a financial institution – or even placing in a trust fund customers' advance payments. The Province of Ontario began to arrange an emergency airlift of some 3,000 travellers stranded in southern sunspots using money from its travel compensation fund (*Globe and Mail*, 28 April 1990, pp. A1–A2). In addition, another 6,000 to 8,000 travellers were due to return the next week. Further, another 8,000 had purchased tickets but had not left on their flights. There was $3.8 million in the compensation fund.
45. p. 253.
46. It is easier to build points with a carrier that flies to all destinations the consumer is interested in. Thus the large carrier may choose to offer one free trip for every thirty paid trips. To offset the difficulty of accruing points since it only flies to a few destinations, the smaller airline may have to provide rewards at a one to fifteen or a one to ten ratio. Tretheway (1989) discusses the nature of these programmes and their success in building brand loyalty.
47. See Canada, Department of Transport (1984).
48. Unfortunately, airline data is scarce in Canada, and when available is published with a two-year lag.
49. The 1980 Canadian figures are likely similar to those in the US in 1977 when discounts were first appearing. By 1980 the US was in full deregulation and discounts were widespread, whereas Canada was only just beginning to allow discounts.
50. On the merger of PWA, then a regional, and CP Air in early 1987, see Gillen *et al.* (1988).
51. Privatisation took place in two stages: one in 1988 and the other in 1989, see section 5.7.
52. See Oum & Gillen, 1983, for the elasticities for Canadian market, and Oum *et al.*, 1986, for the US elasticities.
53. More generally, see Gillen *et al.* (1985) and Stanbury & Tretheway (1987).
54. Christopher Waddell, 'Ottawa to begin selling off Air Canada', *Globe and Mail*, 13 April 1988, pp. A1, A9.
55. For a more detailed critique, see Val Udvartely, 'Why should public buy shares in something it already owns?' *Globe and Mail*, 14 April 1988, p. A7. This particular assessment contains numerous errors of fact and logic, but does assemble in one place all of the bad arguments against privatising Air Canada.
56. 'Critics slam sale of Air Canada', Vancouver *Sun*, 13 April 1988, p. E1.
57. Waddell, *Globe and Mail*, 13 April 1988, p. A9.
58. 'Selling Air Canada', (editorial) *Globe and Mail*, 13 April 1988, p. A6.
59. Of particular importance is the Winnipeg overhaul base with 400 employees. Its existence had been threatened by Air Canada's choice of new aircraft.
60. Note the Alberta legislation still applies to PWA Corp, the holding company of Canadian Airlines International. The new name was adopted after PWA paid $300 million for CP Air early in 1987.
61. On 6 June, the *Financial Times* (p. 3) published a Decima poll which found 11% strongly supporting the plan to privatise Air Canada; 58% supported it; while 21% opposed it and 7% strongly opposed it.
62. *Globe and Mail*, 30 August 1988, p. B1.

63. On the day the merger was announced (19 January) Air Canada's stock rose more than $1 to close at $9. They rose to $9.50 the next day. PWA Corp's shares also rose $2.63 to close at $17.38 on 19 January 1989.
64. *Financial Post*, 2 October 1989, p. 23.
65. Vancouver *Sun*, 6 July 1989, p. C1.
66. *Financial Times*, 10 July 1989, p. 16.
67. On 14 January 1985 Mr Mulroney said that 'Air Canada is not for sale. There may be some persuasive arguments in the case of Air Canada that some people can make in regard to the disposition of equity. I'll take a look at it. But Canada needs a national airline and it's going to have one' (see *Globe and Mail*, 15 January 1985, pp. 1, 2).
68. On 22 August 1987 the *Globe and Mail* (pp. B1, B4) reported that in July the Prime Minister had vetoed the sale of Air Canada 'for fear that his credibility would be further eroded' in light of his statement in January 1985. The sale of Air Canada was deemed to be too politically risky prior to the next general election (*Financial Times*, 7 September 1987, pp. 1–2).
69. On 28 January 1988 the *Globe and Mail* (p. B1) reported that Air Canada had asked the federal government for $300 million in additional equity to finance the renewal of its fleet. This move, the president of Air Canada later admitted, was designed to push the Mulroney Government into privatising the airline. CAIL immediately announced it was opposed to the government supplying Air Canada with more funds (see also *Globe and Mail*, 14 April 1988, pp. B1, B4).
70. While load factors are high, prices are low. The industry has excess capacity, but until it can reduce it, carriers use seat management to lower price and fill as many seats as possible. The economist would say that capacity is being utilized by pricing above variable costs but below total costs. In the long run, capacity will be reduced and prices increased to become compensatory.
71. Wardair did not offer an extensive scheduled service until April 1988 when its new Airbus A-310s began to arrive. The A-310s did not arrive in numbers until mid-1988, then they proved to be *too* large (195 seats). To be more competitive in the full-fare economy (Y) or business class (J) market segments, Wardair needed to offer more *frequent* service (with smaller aircraft). Wardair admitted it should have ordered MD-88s, which seat about 130. The MD-88s were not to begin to arrive until November 1989 – hence they would miss the peak summer period of 1989. In any event, Wardair was too late in making the switch from charter to scheduled service.
72. See M. Drohan, 'Declines in prices of food, transport moderate inflation', *Globe and Mail*, 16 September 1989, p. B1. Domestic future levels in 4th quarter 1989 were little changed from those in the first quarter of 1988.
73. On 6 February 1990 the *Globe and Mail* (pp. B1, B6) reported that PWA Corp would be selling 14 Wardair aircraft (12 A-310s-300s and 2 B-747-100s) for $900 million. They carried $585 million in debt. PWA Corp's stock closed at $12, up 12 cents. The stock was over $17 in January when PWA Corp made its offer to Wardair. Six of the 14 aircraft will leave the fleet in 1990, five in 1991 and three in 1992. The aircraft will be replaced by B-747-4002, B-767-300ERs (Spring 1990) and A-320-200s (expected in 1991). The *Financial Post* (6 February 1990, p. 3) reported that PWA Corp will reap a profit of $120 million over three years on the sale. 'The sale of the planes closes the book on Wardair as a distinct company . . . PWA stopped distinguishing separate Wardair flights last month and started offering all flights under the Canadian Airlines banner'.
    In 1989 CAIL offered 13% fewer seats than in 1988. In 1989, PWA Corp, CAIL's parent, lost $56 million on revenues of $2.67 billion. Air Canada had a profit of $149 million on revenues of $3.68 billion in 1989. In the first quarter of 1990 PWA Corp lost $34.2 million, while Air Canada lost $14 million.
74. See Tretheway (1990b) for a discussion of Canada's international airline policy.
75. See Goldman (1986) and Stanbury (1983) for somewhat different interpretations.

References

Andriulaitis, R. J., Frank, D. L., Oum, T. H. & Tretheway, M. W. (1986) *Deregulation and airline employment: myth versus fact*, Vancouver: Centre for Transportation Studies, University of British Columbia.

Baggaley, C. D. (1981) *The emergence of the regulatory state in Canada, 1890-1939*, Regulation Reference Technical Report No. 15, Ottawa: Economic Council of Canada.

Bailey, E. E., Graham, D. & Kaplan, D. (1985) *Deregulating the airlines*, Cambridge, Mass: MIT Press.

Baldwin, J. R. (1985) *The regulatory agency and the public corporation*, Cambridge, Mass: Ballinger.

Breyer, S. G. & Stein, L. R. (1982) 'Airline deregulation: the anatomy of reform'. In Poole, R. W. (ed.) *Instead of regulation*, Lexington, Mass: D. C. Heath.

Canada, Department of Transport (1981a) *Economic regulation and competition in the domestic air carrier industry*, Ottawa: Department of Transport.

Canada, Department of Transport (1981b) *Proposed domestic air carriers policy (unit toll services)*, Ottawa: Department of Transport.

Canada, Department of Transport (1984) *New Canadian air policy*, Ottawa: Department of Transport.

Canada, Department of Transport (1985) *Freedom to move: a framework for transportation reform*, Catalogue No. T22-69/1-985E, Ottawa: Minister of Supply and Services.

Canada, House of Commons Standing Committee on Transport (1980) *Minutes of proceedings and evidence*, Issue No. 2, 2 December 1980.

Canada, House of Commons Standing Committee on Transport (1982) *Domestic air carrier policy*, Ottawa: Minister of Supply and Services.

Canada, House of Commons Standing Committee on Transport (1985) *Freedom to move: change, choice, challenge*, Sixth Report, December 1985.

Canadian Transport Commission (1984) *Interim report of the Air Transport Committee of the Canadian Transport Commission on domestic charters and airfare issues*, Ottawa: Canadian Transport Commission, Air Transport Committee.

Carlton, D., Landes, W. & Posner, R. (1980) 'Benefits and costs of airline mergers: a case study (market share due to single carrier service in North Central Southern Merger)', *Bell Journal of Economics*, 11(10), 65-83.

Caves, D. W., Christensen, L. R. & Tretheway, M. W. (1984) 'Economies of density versus economies of scale: why trunk and local service airlines costs differ', *Rand Journal of Economics*, 15(4), 471-89.

Caves, D. W., Christensen, L. R., Tretheway, M. W. & Windle, R. J. (1987) 'An assessment of the efficiency effects of U.S. airline deregulation via an international comparison'. In Bailey, E. (ed.) *Public regulation: new perspectives on institutions and policies*, Cambridge, Mass: MIT Press.

Derthick, M. & Quirk, P. J. (1985) *The politics of deregulation*, Washington, DC: Brookings Institution.

Douglas, G. W. & Miller, J. C. (1974) *Economic regulation of domestic air transport: theory and policy*, Washington, DC: Brookings Institution.

Dresner, M., Hadrovic, C. & Tretheway, M. W. (1988) 'The Canada–U.S. air transportation bilateral: will it be freed?' *Proceedings, Canadian Transportation Research Forum 25th Annual Meeting*, University of Saskatchewan Printing Services.

Economic Council of Canada (1981) *Reforming regulation*, Ottawa: Minister of Supply and Services.

Economic Council of Canada (1986) *Minding the public's business*, Ottawa: Minister of Supply and Services.

Ellison, A. P. (1985) 'The new air transport policy: liberalization not deregulation', mimeograph, Economic Council of Canada Government Enterprise Project.

Gillen, D. W. (1979) 'Bill C-3: the new Air Canada Act', In Reschenthaler, G. B. & Roberts, R. (eds) *Perspectives on Canadian airline regulation*, Montreal: The Institute for Research on Public Policy, pp. 193-9.

Gillen, D. W., Oum, T. H. & Tretheway, M. W. (1985) *Canadian airline deregulation and privatization: assessing effects and prospects*, Vancouver: Centre for Transportation Studies, University of British Columbia.

Gillen, D. W., Oum, T. H. & Tretheway, M. W. (1986) *Airline costs and performance: implications for public and industry policies*, Vancouver: Centre for Transportation Studies, University of British Columbia.

Gillen, D. W., Oum, T. H. & Tretheway, M. W. (1987) *Identifying and measuring the impact of government ownership and regulation on airline performance*, Discussion Paper No. 326, Economic Council of Canada.

Gillen, D. W., Stanbury, W. T. & Tretheway, M. W. (1988) 'Duopoly in Canada's airline industry: consequences and policy issues', *Canadian Public Policy*, **14(1)**, 15-31.

Goldman, C. S. (1986) 'The Competition Act as it relates to the regulated sector', speech to the Canadian Association of Members of Public Utility Tribunals, 10 September, mimeograph, Ottawa: Bureau of Competition Policy.

Greig, J. A. (1977) *Regional air carrier safety*, Report #40-77-2, Ottawa: Canadian Transport Commission.

Jordan, W. A. (1970) *Airline regulation in America: effects and imperfections*, Baltimore: Johns Hopkins Press.

Kraft, D. J. H., Oum, T. H.   Tretheway, M. W. (1986) 'Airline seat management', *Proceedings of Canadian Transportation Research Forum 21st Annual Meeting*, Vancouver, BC, May, Saskatoon: University of Saskatchewan Press, pp. 232-45, also in *Logistics and Transportation Review*, **22(2)**, 115-30.

Levine, M. E. (1965) 'Is regulation necessary? California air transportation and national regulatory policy', *Yale Law Journal*, **74(8)**, 1416-47.

Morrison, S. A. & Winston, C. (1986) *The economic effects of airline deregulation*, Washington, DC: Brookings Institution.

Morrison, S. A. & Winston, C. (1989) 'Enhancing the performance of the deregulated air transportation system'. In *Brookings Papers on Economic Activity: Microeconomics*, Baily, M. N. & Winston, C. (eds.), 61-112.

Oum, T. H. & Gillen, D. W. (1983) 'The structure of intercity travel demands in Canada: theory, tests and empirical results', *Transportation Research*, **17B**, 175-91.

Oum, T. H., Gillen, D. W & Noble, D. S. E. (1986) 'Demands for fare classes and pricing in airline markets', *Logistics and Transportation Review*, **22(3)**, 195-222.

Oum, T. H. & Tretheway, M. W. (1984) 'Reforming Canadian airline regulation', *Logistics and Transportation Review*, **20(3)**, 261-84.

Reschenthaler, G. B. & Roberts, B. (eds) (1979) *Perspectives on Canadian airline regulation*, Butterworth & Co. (Canada) for the Institute for Research on Public Policy, Montreal.

Reschenthaler, G. B. & Stanbury, W. T. (1982) 'Canadian airlines and the visible hand', draft book manuscript prepared for the Institute for Research on Public Policy.

Reschenthaler, G. B. & Stanbury, W. T. (1983) 'Deregulating Canada's airlines: grounded by false assumptions', *Canadian Public Policy*, **9(2)**, 210-22.

Stanbury, W. T. (1983) 'Provincial regulation and the Combines Investigation Act: the Jabour case', *Windsor Yearbook of Access to Justice*, **3**, 291-347.

Stanbury, W. T. (1986) 'The New Competition Act and Competition Tribunal Act: "Not with a bang, but a whimper" ', *Canadian Business Law Journal*, **12(1)**, 2-42.

Stanbury, W. T. (1988) 'Privatization and the Mulroney Government'. In Gollner, A. B. & Salee, D. (eds) *Canada under Mulroney*, Montreal: Vehicule Press, pp. 119-57.

Stanbury, W. T. & Tretheway, M. W. (1987) *Analysis of the changes in airline regulation proposed under Bill C-18*, mimeograph, Submission of the House of Commons Standing Committee on Transport.

Tretheway, M. W. (1984) 'An international comparison of airlines', *Proceedings of Canadian Transportation Research Forum 19th Annual Meeting* (Jasper, Alta, May), pp. 653–76.

Tretheway, M. W. (1989) 'Frequent flyer programs: marketing bonanza or anticompetitive tool?' *Proceedings at Canadian Transportation Research Forum 21st Annual Meeting*, Halifax, Nova Scotia, June 1989, Saskatoon: University of Saskatchewan Press, pp. 433–45.

Tretheway, M. W. (1990a) 'Globalization of the airline industry and implications for Canada', to be published in *Proceedings of Canadian Transportation Research Forum 25th Annual Meeting*, Saskatoon, Saskatchewan, Saskatoon: University of Saskatchewan Press.

Tretheway, M. W. (1990b) 'Canada and the changing regime in international air transport', Working Paper 90-TRA-006, Faculty of Commerce and Business Administration, University of British Columbia, Vancouver.

White, L. J. (1979) 'Economies of scale and the questions of "natural monopoly" in the airline industry', *Journal of Air Law and Commerce*, **44**(3).

Index